D0855727

REF   QB   Jaco+
      981
      •J2371
      1986

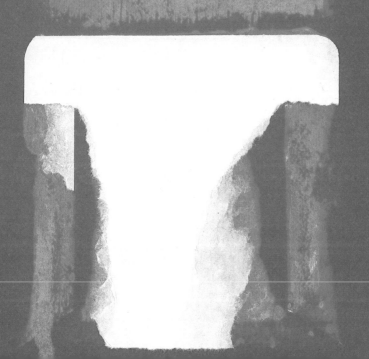

A HERETICAL COSMOLOGY

OTHER BOOKS BY THE SAME AUTHOR

L'univers en marche (*The Universe in Motion*). Imprimerie Messeiller. Neuchâtel (Switzerland) 1944. Republished in Nouvelles éditions latines. Paris 1952.

Idées nouvelles sur l'univers (*New Concepts of the Universe*). Imprimerie Messeiller. Neuchâtel 1951.

Attraction ou distraction universelle? (*Universal Attraction or Distraction*) Nouvelles éditions latines. 1954.

La Terre s'en va (*Earth's Flight Beyond*) Edition de la Table ronde. Paris 1976.

Eléments de physique evolutive (*Elements of Evolutive Physics*). Paris 1962.

Méditations sur le mouvement (*Reflections on Motion*). Paris 1963.

L'évolution universelle (*Universal Evolution*). Paris 1964.

Works on sale through Editions de la Pensée universelle, 4, rue Charlemagne, 75004 Paris.

Histoire critique de la pensée (*Critic History of Thought*). Editions de la Pensée universelle. 1970.

4 volumes sold separately:

Volume I. *La bataille des idées en religion* (*The Battle of Ideas in Religion*)
Volume II. *La bataille des idées en philosophie* (*The Battle of Ideas in Philosophy*)
Volume III. *La bataille des idées en science* (*The Battle of Ideas in Science*)
Volume IV. *L'approche de l'harmonie* (*The Approach of Harmony*)

L'éloge de la sottise (*The Praise of Fools*). La Pensée universelle. Paris 1971.

Préjugés et contradictions de la science (*Preconceptions and Contradictions in Science*). La Pensée universelle. 1973.

L'imposture scientifique (*Scientific Imposture*). La Pensée universelle. Paris 1973.

Translations of the above works:

L'univers en marche. Italian translation. *L'universo in camino*. Published by Satet. Turin.

L'univers en marche. Spanish translation. *El universo y la tierra*. Espasa Calpe Argentina. Buenos Aires.

L'évolution universelle. English translation.
Universal Evolution. Editions du Mont Blanc. Geneva.

Histoire critique de la pensée. Portuguese translation. *Historia critica do pensamento*.
Published by Mundo Musical. São Paulo.

La Terre s'en va, 2nd edition. English translation:
Earth's Flight Beyond. Published by Robert Hale. London 1977.

A HERETICAL COSMOLOGY

The Catastrophic Dislocations
of Galaxies, Stars and Planets

LOUIS JACOT

Profusely Illustrated

AN EXPOSITION-BANNER BOOK

Exposition Press of Florida, Inc.   *Pompano Beach, Florida*

This book has been published in French under the title:

Science
et Bon Sens
Défin à l'Académie des sciences de Paris
qui refuse d'admettre que le système solaire est en expansion
et que les planètes s'éloignent progressivement du Soleil
conformément à la loi de Bode

English translation:

Science
and Common Sense
A Challenge to the Paris Academy of Science
which refuses to admit that the solar system is
expanding and that the planets
are moving progressively away from the Sun
in accordance with Bode's Law

FIRST AMERICAN EDITION

© 1986 by Louis Jacot

All rights reserved. No part of this book may be reproduced, in whole or in part, in any form or by any means, electronic or mechanical, including photocopying, recording, or by any information storage and retrieval system, without permission in writing from the publisher. Address inquiries to Exposition Press of Florida, Inc., 1701 Blount Road, Pompano Beach, FL 33069.

Library of Congress Catalog Card Number: 84-90318

ISBN 0-682-40175-7

Printed in the United States of America

Contents

List of Illustrations

ix

Preface: An Unhappy Marriage

Science and common sense do not make a happy marriage, and there are good reasons underlying their mutual mistrust. For thousands of years common sense told men that the Sun revolves around the Earth. For centuries famous astronomers, inspired by Hipparchus and Ptolemy, the most renowned of all, made all the planets—except for the Earth, the center of the Universe—dance in orbits festooned with epicycles, small superfluous circles derived from their direct observations. In fact, their observations taught them that the planets which revolved around the Sun stopped in their courses from time to time, then went back before starting forward again. In their eyes epicycles could only be the actual paths of the planets since they tallied with their observations. It was Copernicus, inspired by Aristarchus of Samos, who realized that the Earth was also a planet and that the epicycles did not correspond to the observed movements of the planets, but were only apparent motions due to the movement of the observatory, the Earth, which also revolved around the Sun.

Science brings knowledge; it gives names to things and measures the development of phenomena. In order to achieve true knowledge the latter must be interpreted correctly. This is where common sense comes in: this consists of using knowledge in order to understand what is happening by reasoning in the right way. True cognition of the world is only possible if knowledge and common sense are closely interlinked.

But how does one know if one is reasoning correctly? And can common sense guarantee certainty? The world is complex, full of innumerable phenomena with manifold interpretations. Thus

xi

Descartes was at pains to prescribe the essential rules for the guidance of one's mind: specifying that it was necessary to begin by making a "tabula rasa" of all preconceptions, to admit only what one had no reason to doubt, to make more and more thorough analyses, followed by all-embracing syntheses in order to establish whether anything had been omitted and whether the ensuing ideas on a phenomenon agreed with others. The elimination of preconceptions and the collation of ideas through synthesis are the "sine qua non" conditions of an approach to the truth, since if it is difficult to achieve certainty of reasoning in the right direction, one may be certain of being in error when ending up with contradictions.

Modern science, however, is so full of them that they trouble investigators less and less, to the point that the scientists who "know" are much more numerous than those who understand. There are even specialists who know nearly everything about a particular subject but understand nothing about the world in which they live.

When a witness states to the judge that at the time of the crime he was at home sound asleep and that at the same time he was winning the last stage of the "Tour de France" on his bicycle, the judge concludes that the witness is dreaming, lying or talking nonsense. When a physicist asserts that light is a wave phenomenon in which the particles of a tenuous medium, ether, undulate together while remaining on the spot and that at the same time these particles speed through space individually at a speed of 300,000 km/sec, the man of good sense, if he is polite, considers the physicist to be a dreamer. But physicists do not agree. Some acknowledge the wave theory on Mondays, Wednesdays and Fridays, that of the emission of flashing photons on Tuesdays, Thursdays and Saturdays and both theories at the same time on Sundays, with the result that due to the convenience of the two contradictory theories, Sunday physicists are by far the most numerous. An interpretation which transcends the ideas generally accepted by the man in the street becomes superscientific. The idea that one may simultaneously remain on the same spot and travel at great speed with respect to the same system of reference has become so familiar to physicists that they have no intention of abandoning it nor of studying it more thoroughly. On the contrary, they are proud of surpassing themselves, like Nietzsche who, by dint of doing so, reached the heights of unreason.

According to Newton, the Earth attracts the Moon at 1.36 mm/sec, the latter progressing along its orbit at a rate of 1.02 km/sec, which causes it to describe an orbit around the Earth. But during this second, the Earth has traveled 30 km, and the Moon, like an obedient offspring, has followed. The man of common sense who has never seen a cyclist crawling along at one kilometer an hour being able to ride for very long around another traveling at a speed of 30 km/h, wonders by what miracle the Moon was able to follow. The same problem arises with regard to the Earth which revolves around the Sun at a rate of 30 km/sec, while the Sun speeds through space at 250 km/sec.

There are so many contradictions in science—from the atomic nucleus to the galaxies—that, in order to get out of the difficulty, some physicists have preferred to avoid the issue by denying the existence of a cause and effect relationship and by constructing a system of quantum mechanics solely upon statistics. But can numbers make any sense if words do not. By denying the relationship of cause to effect aren't we denying the existence of Time, master of the world, and removing any sense from scientific research, including statistics?

Even certain biologists have followed this path, and one of the most famous of them has written a book with philosophical pretensions entitled *Le hasard et la nécessité* (*Chance and Necessity*) in which he attempts to demonstrate that there is no intelligence in the world. It is true that no biologist has ever discovered its presence in the form of grains or particles, even under an electronic microscope. No more would it be discovered in a computer or a supersonic aircraft. But does this mean that as far as a computer is concerned that it may be assembled by chance? To deny intelligence to Nature is to deny it to Man who is an integral part of it, which implies that this author has written a book in which the letters and words succeed one another by chance. We would not go this far.

Is it really impossible to reconcile science and common sense? Why not start by setting out the fundamental problems correctly, then compare the opinions expressed and eliminate preconceptions by going back to the causes of contradictions?

Is it really against all the odds to attempt this reconciliation and to obtain a true picture of our world and its development?

1

The Fundamental Problems Concerning the Structure and Evolution of the Universe

THE PROBLEMS POSED BY THE PHILOSOPHERS OF ANCIENT GREECE

From earliest antiquity men have recognized the existence of a world of invisible forces acting on the visible world of appearances and have tried to find out which was the real world, that of causative action or that of visible effects.

In contemplating the variety of bodies and the changes brought about by Nature, the Ionians, perhaps influenced by Indian thinkers, considered that everything that existed resulted from the combination of a single universal substance. This idea springs from profound consideration of the invisible world of forces which in their eyes represented the real essence of things. Their persistency in searching for the universal substance is a mark of genius. Although their works are lost, several fragments indicate that for Thales of Miletus this primordial element was water; for Heraclitus of Ephesus, fire; for Anaximander, ether; while for Anaximenes it was air.

In adopting water as the primordial element, Thales was not very far from the modern materialists since water (H_2O) is composed of one atom of oxygen and two atoms of hydrogen, the simplest element of Mendeleev's classification. But such considerations do not

1

appear to have led Thales to make this choice. Convinced that everything evolves and that life came out of the water, it was rather this idea of the evolution of living creatures from an aquatic environment which probably led Thales to this concept.

For Anaximander of Miletus, the primordial element had to be more tenuous than water. He adopted ether which he defined as being unique, infinite-indefinite and capable of movement. By closely blending together substance and movement, Anaximander gave ether an inspired definition.

A follower of Anaximander, Anaximenes thought that the primordial element was air; probably because according to the Hindu thinkers, by whom he was influenced, air—atman—is the breath of life, that which we breathe, which animates all living beings and maintains the atman-Brahman contact, this mystical union of the individual soul with that of the Universe.

Although some of Heraclitus' theories are not very lucid— Socrates called him "Heraclitus the obscure"—the idea that everything is movement and that everything changes is expressed unequivocally in certain sayings which have been preserved for us, such as "one never bathes twice in the same stream" or "nothing is, everything is becoming." This idea of universal evolution is the very foundation of his thinking, so that in the *Timaeus* Plato defines science as "a credible prediction of future development." If Heraclitus regarded fire as the primordial element it is because fire is a manifestation of the invisible element which causes change. It is enough to replace the term "fire" by that of "energy," the meaning of which is not too far removed, in order to arrive at a perfectly modern concept.

The influence of Indian thought is especially felt with Pythagoras who considered that in a world where there is only one universal substance, the essence of quality could only reside in numerical relationships, proportions and combinations. As far as Pythagoras was concerned, numbers constituted the essential principles even of things. The Cosmos is a well-ordered whole whose parts are joined to one another by bonds of harmony, numbers being the eternal principles upon which this harmony rests. The influence of Pythagoras was all the greater due to the Persian invasion which forced him to leave Samos, his island birthplace, to settle in Italy where he founded a school of philosophy, politics and religion, the Pythagoreans never making any distinction between science, wisdom and religion. Our modern chemists who adopt Mendeleev's classification of the elements according to the number of their constituent particles and their distribution within the nucleus and around the periphery, are following in the Pythagorean tradition.

Probably influenced by the Pythagoreans, the Eleatic philosophers did not agree with the Ionians regarding the importance of movement. In the eyes of Parmenides, movement hides the reality of "Being," eternal and immovable. Changes are temporary and cannot affect the Whole whose unity and eternity are essential qualities. The fundamental idea of the unity of being caused him to declare that "Being is, and non-being is not," which appears to be an obvious truism, but is derived from profound considerations which not only exclude nothingness and imply the continuum of the universe, but give priority to a spiritual conception of the world.

Zeno of Elea, a pupil of Parmenides, developed the ideas of his master with regard to one and many and showed that something cannot be divided to infinity, for if its constituent parts still have a magnitude, their total is infinite, while if their magnitude is zero, their total is zero.

Inspired by Zeno's demonstration, Leucippus and Democritus of Abdera arrived at the notion of indivisible atoms, a notion which would be taken up again by modern physicists in a form which moreover evolved.

Extracting the quintessence of the ideas expressed by the Ionians, the Pythagoreans and the Eleatics, Anaxagoras of Clazomene maintained that "everything is in everything." There is only one universal substance containing all the elements necessary for the make-up of existing entities. Probably also influenced by Indian thinkers and their conception of the soul of the world was well as by the Egyptians for whom creation consisted of the ordering of matter, "Khet," by the universal spirit, "Ka," which thus created individual souls, "Ba," Anaxagoras considered that the prime mover which brought order to the world was "Nous," the organizing intelligence which ruled the universe. This essentially spiritual conception of the world did not, however, prevent Anaxagoras from regarding the universe as a great vortex. According to him, the Sun, the Moon and all the stars were only incandescent rocks drawn along by the rotation of the ether; this was considered an impiety by the Athenians who forced him to leave Athens despite the protection of Pericles.

Though Anaxagoras thought that everything is in everything, he nevertheless acknowledged that "Nous" was of an essence superior to the rest of the world, which seems to have given him some trouble in explaining the relationship between "Nous" and the rest of the world. Socrates also, fired with enthusiasm by the concept of "Nous," was greatly deluded by the influences drawn by Anaxagoras.

However, this superiority of the invisible over the visible world and of the spiritual over the material prompted Socrates and his disciple Plato to formulate a complete conceptual system of the world of ideas.

To be sure, among the Greeks, each school developed its own ideas of the world. Some thinkers, like Empedocles of Agrigentum, even adopted several elements—earth, water, air and fire—as fundamental constituents, but the majority were proponents of the single substance. In their efforts to comprehend the world, the Greeks thus came to the conclusion that the universe formed a unity governed by a principle of harmony which gave it its underlying meaning, even if Heraclitus, for example, assumed that harmony could only be achieved by the conflict of opposites. Neither gods nor men were outside the universe, but formed an integral part of it just as other things did. Matter, interstellar spaces, life, mind and all phenomena formed the inseparable parts of a great whole.

While the majority of the ancient Greeks placed the Earth at the center of the universe, Aristarchus of Samos, in the third century before Christ, apprehended and taught that it revolved around the Sun. His heliocentric system was, however, not adopted since it was too advanced, even for astronomers as famous as Hipparchus and Ptolemy.

In short, the preoccupations of the most eminent Greek philosophers bore upon the fundamental problem dominating the universe, in particular concentrating on the relations between the real world of forces and that of appearances. If they did not succeed in solving the problems raised, they had the signal merit of posing them correctly in the following forms:

1. Is there one universal substance or are there several?
2. If there are several of them, what are they and how are they related?
3. If there is only one, what is its nature and how can it form different entities?
4. How are substance and movement related?
5. What are the relationships between the whole and the parts?
6. What are the movements of the heavenly bodies and how are they related to interstellar space?
7. How are the material and spiritual related?

These problems confront us now as crudely as they did more than two millennia ago.

THE CARTESIAN VIEW OF THE UNIVERSE

It was as a result of being inspired by the ancients that the men of the Renaissance broached the great problems of the universe. Copernicus, who was acquainted with Aristarchus' theory, referred to it in his celebrated work *De revolutionibus orbium coelestium*, which he published the year of his death in 1543. This theory incurred the wrath of the Church which condemned its adherents. Giordano Bruno was burned alive in 1600 and in 1633 Galileo was forced to retract this "absurd and heretical doctrine." This fearful pressure exercised on men's minds by the Church had the effect of imposing the Judaic conception of a clear separation between God and the world, and of limiting the questions capable of being discussed by excluding in particular everything spiritual, which was permanently shrouded in dogma.

Thus curtailed, the universe which had formerly consisted of an active, vibrant and thinking whole was reduced to the mechanics of matter and interstellar space. This curtailment compelled thinkers to concentrate their efforts on the problems tolerated by the Church. With the aid of instruments, Galileo discovered the satellites of Jupiter, and many researchers began to study the phenomenon of light, knowledge of which they considered necessary to the solution of the problem of the structure of the universe. However, it would be wrong to assume that the study of these phenomena prompted thinkers right away to make a fundamental distinction between the matter forming the heavenly bodies and interstellar space.

Hence Descartes (1596-1650), who adopted the concept of the ancients according to which "the universe abhors a vacuum," considered it as being continuous and formed of a single matter, whose consistency is very variable. He assumed that between the Sun and an observer on Earth, space consists of a thin matter in the form of small spheres bunched together like lead shot on which the particles escaping from the Sun exercise a "vibrating force" which recurs and produces the phenomenon of light. Thus Descartes regarded the ether as material acting as a transmitting medium. These small spheres which collide with one another not only have the effect of causing vibration, but are actuated by an overall movement which causes them to describe huge vortices drawing the stars along with them while remaining in contact with each other. Descartes made many drawings of these juxtaposed vortices. This is a rough summary of Descartes' theory although it is impossible to say whether it accurately reflects his thinking, since becoming more prudent in the wake of the fates which had befallen Giordano Bruno and Galileo, he knew that he was watched by the Church, even in Holland. Pascal, moreover, reproached him for having had recourse to God only for the initial flick of the finger to start things off.

The main elements of Descartes' theory are as follows:

1. The continuity of the universe.

2. The unity of the universal matter.

3. The movements of this universal matter whose main effects are to transmit light phenomena and form large vortices.

Since the physicists have called bodies composed of atoms "matter," this term employed by Descartes seems wrong when it applies to interstellar space. But Descartes speaks of "fine" matter. Spinoza defined the term "substance" as everything which exists in itself. The concept of small spheres colliding with one another like lead shot may seem simplistic, but does this differ so much from that of Planck with regard to quanta "which behave like actual *atoms of substance* in their collision with matter" (Max Planck)?

Malebranche (1638–1715), a follower of Descartes, made a closer study of this ether of interstellar space and taught that the cohesion of heavenly and material bodies was due to the pressure of this invisible and restless matter, itself composed of elastic spheres which may form from small vortices.

Malebranche thus rounded out Descartes' theory on two important points:

1. The universal substance may, by its pressure, form material and heavenly bodies.

2. Small vortices may be formed within elastic spheres.

The discovery in the twentieth century of the "spin" or internal movement of particles would prove the accuracy of Malebranche's intuition.

THE NEWTONIAN VIEW OF THE UNIVERSE

In his *Philosophiae naturalis principia mathematica*, published in 1687, Newton (1643–1727) seems to have taken the opposition to Cartesian rationalism as his first principle. Already in 1672 in his treatise on light, he stated that the latter is formed of particles of luminous essence which travel through space like rifle bullets. In his attempt to express his world system in mathematical form, he set forth the law which made him famous, according to which "everything occurs as though bodies are attracted to one another in proportion to their mass and in inverse ratio to the square of their distance."

Newton thus completely overturned the generally accepted concepts regarding the nature of light and the movement of the heavenly bodies. According to Descartes, Malebranche and other contemporaries of Newton, such as Hooke, Huygens and Leibniz, light is a vibratory phenomenon of ether, this substance which draws stars along in their courses and compresses them. According to Newton, both light "corpuscles" and the stars each have their particular motive force which pushes them through empty space. The stars even appear to have two motive forces, one behind which pushes them forward, the other in front which attracts other heavenly bodies, unless they have multidirectional motors which push in one direction and simultaneously draw in all directions by interlocking atoms.

The Newtonian view implies the emptiness of interstellar space, otherwise the stars would be slowed down by the obstacles they encounter. Thus the ether must disappear. Oh no it must not! This is not Newton's idea after all when he writes to Bentley that one must be completely bereft of all sense of philosophy to believe that the attraction of the stars may be transmitted in the absence of a transmitting medium. So ether is necessary to Newton's theory, but the celebrated formula does not take this into account. In it the ether is equal to zero. There is thus a divorce between the theory and its mathematical expression, which is scarcely compatible with the spirit of philosophy.

Having died in 1650, Descartes could hardly refute these published in 1672 and 1687. But other eminent thinkers took it upon themselves, in particular Hooke (1635–1703), Huygens (1629–1695) and Leibniz (1646–1716).

This broaching of the fundamental problem of the nature of light and the structure of the universe is of primary importance. It is essential that we give it our full attention.

In 1662 Hooke, an able experimenter, was appointed "curator of experiments" of the Royal Society. He carried out a number of experiments on light phenomena and observed the interferences which led him to uphold the wave theory of light against Newton. In his view, the agitation of the medium was due to uniform pulsations perpendicular to the direction of propagation. He displayed extraordinary insight in his study of the relationships between matter and movement which he considered to be two primordial realities of the same nature, each capable of substituting for the other. Thus, in the middle of the seventeenth century, he predicated *the equivalence of mass and energy, but in a much wider form than Einstein, asserting that interstellar space also has mass since it is capable of movement.*

According to Huygens, the ether, the site of the vibratory movements, had to be material since it set material substances in motion. Hence in the seventeenth century he gave the word "matter" a much wider meaning than today, and quite logically so.

Huygens also established the laws of falling bodies, stated the principle of the conservation of momentum and in 1673 published the first treatise on dynamics (*Horologium oscillatorum*). He also investigated the law of centrifugal force in circular motion which he applied to the study of gravity and its variations over the Earth's surface. Furthermore, he was opposed to Newton's natural attractive force, which he regarded as an absurdity.

Leibniz, a thinker of great breadth of mind, went beyond Cartesian philosophy by attributing to things not only dimensions, but momentum which, according to the laws of Galileo, is proportional to the square of their speed. This momentum is not at all the *vis attractiva* of Newton, and in his *Antibarbarus physicus*, Leibniz criticizes the resurrection of the scholastic qualities and chimerical powers by which bodies are attracted by one another like lovers.

As for Malebranche, he wrote that thinkers would make themselves appear ridiculous if they were to attribute attractive powers to oxen in order to explain how wagons followed them. It is by treading the ground with its feet and via the yoke connected to the wagon that the ox draws the vehicle forward.

Newton was hard put to reply to these criticisms which rested upon objective observation of the facts. First he needed either to explain attraction since he did not accept that a force could act at a distance without a transmitting medium. But this ether greatly troubled him. Then the natural force of attraction seemed to him to be questionable, since he did not really see how it acted. He thought he saw forces of this type in magnetism and electricity and assumed that they might intervene in gravity. But as he was unable to explain *how* attraction worked, he states in *"Opticks"* that he is not concerned with the *cause* of these attractions. *He intended to consider these forces mathematically rather than physically.* Thus he abandoned the sphere of philosophy, which according to Aristotle is the search for causes, to restrict himself to mathematics. He renounced getting to the root of things; instead he was content with calculations by way of explanation, thus giving priority to mathematics over logic.

EINSTEIN'S VIEW OF THE UNIVERSE

THEORIES ON LIGHT

An appreciation of Einstein's theories first requires an examination of the theories of light.

On the death of Newton, the scientific world found itself faced with two currents of thought. According to the followers of Descartes, Malebranche, Hooke, Huygens and Leibniz, light is a wave phenomenon due to material impact on the ether which fills interstellar space and transmits the wave by an oscillation on the spot of the particles of this ether. In their view, the universe is continuous.

According to Newton's adherents, the phenomenon of light results from the actual displacement of particles of luminous essence, photons, through empty space. The universe is discontinuous.

As early as 1821 Frauenhofer succeeded in measuring the length of various waves through the development of spectroscopy. Studying characteristic spectral lines and their different positions in spectra, Doppler and Fizeau compared light phenomena to those of sound and reached the conclusion that the light source was moving closer when displacement occurred towards the violet, and moving farther away when the line was nearer to the red. Furthermore, Fizeau corroborated Fresnel's experiments and established that the speed of light was not the same in water as in air, that it increased when the *transmitting* medium was in motion in the light phenomenon's propagation direction, and diminished when the medium was moving in the opposite direction from this propagation.

In discovering induced currents in 1831, Faraday proved on the one hand that there was no long-range action through a vacuum, nor were there instantaneous actions, but that there was always a transmitting medium, and that electrical and magnetic effects were transmitted by the dielectric, a medium capable of being polarized. For him also, the ether was, in a way, a dielectric-like material media.

Then in 1864 Maxwell put forward his theory regarding the electromagnetic nature of ether, according to which the latter is not an inert medium, but appears to be a receptacle of energy. After various measurements, he announced that electromagnetic disturbances travel at the same speed as the phenomenon of light. This finding prompted him to conclude that the medium transmitting light and electromagnetic vibrations was identical for both.

The final step was taken by Hertz who in 1885 succeeded in producing waves of one meter. These waves, which displayed the same phenomena of reflection and refraction as light, were propagated at the same speed as the latter. With the knowledge of radio waves and their practical applications (radio, television, etc.), the existence of ether, its capacity to undulate and its nature can no longer be held in doubt.

The fantastic speed of light—300,000 km/sec—is of the same order as that of electricity in good conductors. This analogy allows us to delve more thoroughly into the problem which this speed poses. The transmission of the phenomenon of light is not comparable to that of the impact which occurs during train shunting when one wagon collides with a second which transmits the impact to a third, which hits a fourth, etc. It is well known that mechanical impact is transmitted more slowly than an electric current. The manufacturers of anti-tank weapons know that a shell which strikes armor-plating at an angle ricochets if the firing of the charge is controlled by a metal pin which sets off the explosive when it hits an obstacle. On the other hand, if the transmission of the impact is carried out electrically, the shell explodes as soon as it touches the tank and does not ricochet. Why this difference? In the first case, purely mechanical transmission occurs by particles knocking into one another, the one receiving the impact first transmitting it to that nearest to it, which reacts to the impact via its elasticity, this in turn delaying transmission. When this occurs electrically, it is the internal movement of the particles which takes place, at the speed of 300,000 km/sec. Electricity is transmitted this rapidly because electrons are perpetually in motion. The current causes a simple acceleration of this motion. In the case of light, the speed of transmission is equally determined by the internal movement of the particles of ether.

A DRAMATIC TURN OF EVENTS: MICHELSON'S EXPERIMENT

Nevertheless the very existence of ether and therefore the wave theory were challenged after Michelson's experiments. Convinced of the existence of this ether, Michelson (1852–1931) tried in 1881 to prove it on the basis of interferences which he thought would occur by emitting two light rays, one in the direction of propagation of the Earth along its orbit, the other in the perpendicular direction. Michelson assumed that the first ray would have a speed of 300,000 km/sec (the speed of light) + 30 km/sec (the speed of the Earth along its orbit), i.e., a total of 300,030 km/sec, while the ray sent out perpendicularly would only have a speed of 300,000 km/sec (solely the speed of light). By bringing back these two light beams into his interferometer, Michelson expected to see interference patterns, which did not happen. In 1887 he repeated his experiment with Morley, again with a totally negative result. The astonishment was general since Michelson's interferometer was of such accuracy that Michelson and Morley published a paper on the possibility of adopting a light wave length as the fundamental standard of length, an idea which was adopted by the General Conference on Weights and Measures in its definition of the meter currently in force.

This negative result, which appeared to mean that 300,000 + 30 = 300,000, astounded the physicists and threw scientific concepts into confusion. In 1893 Fitzgerald put forward the hypothesis of an effective contraction of lengths in the direction of movement. This idea was taken up again by Lorentz in 1903, the latter author accepting the thesis of an immobile ether.

By considering two systems of reference, one immobile and linked to the ether (absolute space and time), the other connected to matter propelled with a uniform motion, Lorentz stated his "transformation formulae," replacing absolute time by local time. The Lorentz transformation thus established a principle of restricted relativity in the field of optics without, however, providing any explanation. It was at this point that Einstein intervened by assuming as a fundamental principle that the speed of light is a constant. This was contrary to the facts ascertained by the experiments of Fresnel and Fizeau which established unequivocally that the speed of light varied according to the transmitting medium and the direction of flow when this medium was in motion. Michelson himself, who had measured this speed a number of times, had established small differences, i.e., a slight diminution by comparison with preceding experiments (we discussed this in *La Terre s'en va—Earth's Flight Beyond*). Therefore it is without any "a priori" justification whatsoever that Einstein laid down the constancy of the speed of light as a basic principle. In order to avoid repetition, we shall analyze later on his theory of special (restricted) relativity and then that of general relativity, when we will examine to what extent Einstein's theory can account for the fundamental phenomena of the solar system.

First, we are going to quickly review the latest discoveries made concerning the solar system, the stars and the galaxies, since it is essential to be acquainted with them in order to come face to face with the theories and assess them in full knowledge of the facts.

2

The Solar System, the Stars and the Galaxies

THE SUN

According to the measurements made by astronomers, the Sun, which has a diameter of 1,392,000 km (109 Earth diameters), is located in a galaxy (the Milky Way) at a distance of approximately 30,000 light years from the galactic center, around which it revolves at a speed of 250 km/sec, which means that it makes a complete revolution approximately every 250 million years. It therefore occupies a rather eccentric position within the galaxy. It revolves on its own axis approximately every 25 days, its rotation being faster at the equator than at the regions close to the poles. In fact its rotation takes 26 days at latitude 30° and 30 days at 60°. Sun spots appear in the belts close to the equator (between $+40°$ and $-40°$) with a maximum occurrence about every eleven years. Devastating solar flares occurring in these centers of activity liberate enormous quantities of energy. Some explosive eruptions create a shock wave which travels at nearly 1000 km/sec.

In addition, the Sun is continuously sending out a magneto-hydrodynamic flow known as the "solar wind," consisting of ionized particles permanently escaping from the corona. Not only electrons, but protons and atomic nuclei are expelled at the time of these eruptions. On reaching the Earth, 149 million kilometers from the Sun, the solar wind is traveling at a speed of 350 to 400 km/sec. It

14

makes its presence known in a particularly spectacular fashion during the passage of a comet, by driving along the debris and dust of the comet away from the Sun, thus producing magnificent cometary tails. The Sun not only travels along its orbit at a rate of 250 km/sec, but also at 20 km/sec in relation to certain stars in the direction of its conventional apex, towards the constellation of Hercules according to some astronomers, towards Vega according to others. Like all radiating stars subject to eruptions, the Sun is continually losing a not inconsiderable part of its mass.

THE PLANETS

The Sun is surrounded by a suite of planets which are by no means randomly distributed through the heavens. In actual fact, the German astronomer Bode found that from Mercury to Uranus, their distances apart were in accordance with a geometrical progression with a common ratio of 2, i.e., 1, 2, 4, 8, 16, 32, 64. There are two exceptions, one at the beginning with regard to Mercury, the other at the end with respect to the planets located beyond Uranus. Of primary importance to an understanding of the solar system, this law will be examined farther on. First we must review the various planets, since each of them has its own particular characteristics peculiar to its stage of evolution.

MERCURY

The nearest planet to the Sun and very dense, Mercury (diameter 4847 km, i.e., 0.38 of the Earth's diameter) is located at a mean distance of 58 million km, but since its orbit is very elongated, it is 46 million km from the Sun at its perihelion and 70 million km at its aphelion. Its speed of translation which is 57 km/sec at the perihelion, falls away to 39 km/sec at the aphelion. Its period of revolution around the Sun is 88 days. Its period of rotation on its own axis is not known for certain; some astronomers estimate it at 58 days, while others think that Mercury always presents the same face to the Sun. Its temperature is between 300° and 400°C in the direction of the Sun—according to distance—while on the opposite side it is below zero. Mercury does not have a satellite. The secular displacement of the aphelion and perihelion of Mercury poses an enormous headache for the Newtonians. Leverrier attempted to attribute it to an unknown planet situated between Mercury and the Sun, whilst Einstein tried to explain it by relativity. Mercury's surface is pitted with craters like the Moon. Some have diameters of over one hundred kilometers.

VENUS

Having diameter of 12,249 km (0.96 of the Earth's diameter), Venus is 108 million km from the Sun and revolves around it in approximately 225 days at a speed of 35 km/sec. The probes sent to Venus revealed a temperature of 750 °K (or 475 °C) at the surface and a pressure of 95 atmospheres (that is 95 times as great as the Earth's surface). The stratosphere revolves 60 times faster than the planet, i.e., in 4 to 5 days, while the planet rotates on its own axis in 243 days, also in a so-called retrograde direction (clockwise). *Venus 8* recorded the most violent winds at an altitude of 50 km (with a speed greater than 100 m/sec). This speed diminishes rapidly towards the ground and becomes less than 1 m/sec below 10 kilometers. The atmosphere of Venus is very dense and consists basically of carbon dioxide. The relief of Venus is very different from the Earth's. The Venusian crust seems to be formed of a remarkably flat, single tectonic plate, most areas having an altitude within a range of 1000 meters. The sole exception to this is Mount Maxwell with an altitude of 11,000 meters, markedly higher than that of Everest.

THE EARTH

Everyone knows that the Earth is a planet, but very few people ask themselves what this means, why it rotates and what the effects of its rotation are apart from the succession of day and night. Its diameter is 12,756 km at the equator and 12,713 km between the poles. This difference of 43 km represents a flattening of 1/297. It revolves around the Sun at a speed of 30 km/sec along an orbit located at a mean distance of 149,600,000 km (April 3 and October 1), a distance which varies between 147,100,000 km at the perihelion (January 2) and 152,100,000 km at the aphelion (July 2). Contrary to what one might think in the northern hemisphere, the Earth is closer to the Sun in winter than in summer. This is why the seasons are more extreme in the southern hemisphere. The Earth rotates on its axis in approximately 24 hours. However, astronomical observations have revealed that the speed of rotation is not constant. Surprising as it may seem, it is greater at the aphelion (the farthest point of its orbit) than at the perihelion. In order that there may be agreement with observations, since the month of January 1972 a second is added at the end of each year.

The axis of rotation which is aligned in the same absolute direction remains parallel to itself during a translation. As it is not perpendicular to the plane of the ecliptic, but inclined at 66° 33', this difference produces seasons which are increasingly pronounced as one approaches the poles where each has light for 6 months and is in darkness for the rest of the year. The Earth's axis of rotation, when hypothetically extended, marks the celestial pole around which the starry heavens appear to revolve in the opposite direction to the Earth's rotational movement. The star closest to this point is currently the Pole Star, but this point shifts with time.

Our calendar year does not correspond exactly to a complete revolution of the Earth around the Sun, i.e., to the sidereal year between two successive passages of a hypothetical line drawn from the Sun to a star which is regarded as fixed. This sidereal year has a duration of 365 days, 6 hours, 9.5 seconds, while the tropical year of our calendar which corresponds to the return of the seasons (successive vernal equinoxes) is shorter by 20 minutes. Furthermore, the farthest point of the orbit, the aphelion, is never in the same position from one year to the next, but advances in the direction of travel, with the result that the year separating two transits at the aphelion is 4 minutes and 45 seconds longer than the sidereal year. Being unable to explain this displacement, the astronomers have called this year the "anomalistic" year.

Gravity is not the same at the equator as the poles, and anomalies have also been established on the northern faces of mountains where, at the same altitude, it is less than that on southern faces.

Seventy percent of the Earth's surface is covered by oceans and 30 percent by continents. The Earth's crust is not compact but is formed of plates which move in relation to one another. They draw apart or approach one another, producing earthquakes and volcanic eruptions. The density of water being 1 and that of surface rocks 2.54, this difference cannot fail to have an influence on the Earth's rotational axis. For the most part, Earth's mountains form chains generally resulting from horizontal and not vertical displacement, as in the case of volcanoes.

It is generally assumed that there is a core at the Earth's center whose diameter would be approximately 6000 km. At approximately 3000 km from the center this is succeeded by the mantle which, on account of certain variations in the propagation of seismic waves, is usually divided into two zones: the upper mantle (approximately 700 km thick) and the lower mantle. The dense mantle, approximately 3000 km thick, is succeeded by the 200-km-thick lithosphere, regarded as virtually stable, on which the unstable crust rests. The latter has a thickness of about ten kilometers under the ocean bottoms and between 35 and 70 km under the continents. The zone upon which the crustal plates slide is generally called the asthenosphere.

Above the Earth's surface is the atmosphere, reaching up to approximately 10 km, which is succeeded by the stratosphere up to 25 km, from 25 to 80 km by the mesosphere, then by the thermosphere whose temperature is very high due to the absorption of ultraviolet rays and X rays.

The circumference of the Earth's globe being slightly greater than 40,000 km, the speed of rotation at the equator is therefore about 1700 km/h.

At 150 km, the gaseous particles have a super-rotation 40 percent greater than that of the planet. At a distance of 15 Earth radii, the speed of co-rotation is 21,600 km/h (or 6 km/sec).

The Earth has a magnetic field which is subject to significant variations in intensity and inclination. Through the ages numerous inversions of the magnetic field have succeeded one another. Beyond the atmosphere there exists a vast plasma, the magnetosphere, comprising a population of particles with diverse characteristics and circulating in all directions. There are belts of charged energy particles known as the Van Allen belts. The lower belt, consisting basically of protons, is situated at a height of less than 2 Earth radii. The outer belt, formed of electrons, is located at between 3 and 4 Earth radii. The magnetosphere extends up to about 10 Earth radii in the direction of the Sun where it is kept back by the solar wind, causing a shock wave, whilst on the side opposite the Sun it extends much farther (at least 500 Earth radii). The magnetosphere is surrounded by a layer where the magnetic field is canceled out, the magnetopause.

At the time of solar flares, charged particles, of very high energy (100 million electron-volts and more), reach the Earth in a few hours. These are solar cosmic rays. Charged particles with lower energies form cosmic showers which move at a speed of 1000 to 2000 km/sec. Their magnetic charge disturbs the Earth's magnetosphere approximately 48 hours after the eruption and this disturbance is felt on Earth as a magnetic storm which often makes radio equipment useless for the duration of the storm.

Since the Earth is a planet, geology is, in fact, a chapter in astronomy, a very exciting chapter which affords us a broad outline picture of the past and from which we learn that through the ages the Earth has undergone very large climatic variations, with glacial periods succeeding tropical climates in regions which today are temperate. Tropical climates even extended very close to the poles. Thus, at the start of the Tertiary, palm trees grew in Alaska. The last glaciation ended some 10,000 years ago. These periodic climatic changes are recorded in rock formations that the geologists have classified into four eras: Primary, Secondary, Tertiary and Quaternary, each being subdivided into numerous distinct layers, distinguished one from another especially by their fauna and flora whose evolution has been able to be reconstructed.

THE MOON

The Earth is the first planet out from the Sun which has a satellite, the Moon, whose diameter is 3,473 km and whose volume is 1/50th that of the Earth. Approximately 60 Earth radii away from the latter, or an average of 384,000 km, it is 356,000 km away at its perigee (the nearest point to the Earth) and 406,000 km away at its apogee (the farthest point). Due to this difference of 50,000 km, the Moon does not always appear to us to be the same size. It revolves around the Earth in 27 days, 7 hours, 43 minutes (sidereal revolution) while the synodic period (interval of time required for the Sun, the Earth and the Moon to return to the same positions— the full moon, for example) is 29 days, 12 hours, 44 minutes. The Moon travels along its orbit at a rate of 1.02 km/sec. The time separating two passages of the Moon is on average 24 hours, 50 minutes. According to the data on eclipses in ancient times, it may be concluded that the Moon was then closer to the Earth and revolved around it more rapidly. This progressive distancing continues up to the present day.

The Moon, which revolves around the Earth in the same direction as the Earth's rotation (in the so-called positive direction, despite it being anti-clockwise), always shows the same face. Its surface is pitted with craters, hollows and rills.

Three hypotheses have been put forward regarding the origin of the Moon:

1. The Moon was formed elsewhere and has been captured by attraction.
2. The Moon condensed independently of the Earth, but in close proximity.
3. The Moon was expelled by the Earth in prehistoric times.

MARS

Mars is situated at approximately 227 million km from the Sun (206 million at the perihelion, 248 million at the aphelion). It revolves around the Sun in about 688 days at an average speed of 24 km/sec. It has a diameter of 6748 km (0.53 of the Earth's), and its period of rotation, in the positive direction like the Earth, is 24 hours, 37 minutes.

The present very low temperature of Mars precludes ice from melting, but the large "canals" bear testimony to the past existence of water in liquid form and, therefore, a warmer climate.

The plateau of Tharsis, in the equatorial region, looks like a bulging-out of the Martian crust. It is here in the neighborhood of this plateau that the largest volcanoes are found. The highest in the region, and probably in the whole solar system, called Olympus Mons, is 500 km across and 24 km high (nearly three times as high as the Himalayas). The country flanking this volcano is cracked and scored by channels which seem to be scars from older volcanic activity.

Mars has two satellites: Phobos (diameter only 22 km) at a distance of 9380 km, which has the peculiarity of revolving around the planet in 7 hours, 39 minutes, or three times faster than the planet's own rotation (24 hours, 37 minutes). Its surface is covered in grooves, attesting to its contact with the planet. The other satellite, Deimos (even smaller, diameter 11 km), situated at a distance of 23,480 km, revolves around Mars in 1 day, 6 hours, 18 minutes.

THE ASTEROIDS

Between Mars and Jupiter, at a mean distance of 410 million km from the Sun, is found not a planet, but a multitude of debris revolving around the Sun in various orbits. These are called the asteroids, some of which have diameters of several hundred kilometers, while thousands of others are of very small dimensions.

In 1804, the German astronomer, Olbers, put forward the theory that the asteroids were the debris from a planet that had disintegrated.

JUPITER

Jupiter, the largest of the planets, has a diameter of 142,880 km (11.2 times the Earth's) and is situated at a mean distance of 777 million km from the Sun (738 million km at the perihelion, 803 million km at the aphelion). It makes a revolution around the Sun in 11 years and 314 days (at a speed of 13 km/sec) and rotates on its own axis faster than the Earth, that is, in less than 10 hours. Gaseous for the most part, it rotates at different speeds: 9 hours, 50 minutes at the equator and 9 hours, 56 minutes near the poles. Its fast rotation causes a significant flattening of 1/15.

At the magnetopause (at 95 Jovian radii), the speed of some particles is of the order of 1000 km/sec, a speed greater than that of the solar wind (400 km/sec), which could mean that Jupiter emits its own "wind." The *Voyager 2* probe discovered several satellites, bringing their number up to 14. In addition, a dust ring has also been found around Jupiter. All the satellites are very different from one another. Ganymede with its 5300 km diameter is the largest satellite in the solar system.

One peculiarity which should be noted is that all the known satellites close to the planet revolve in the positive direction, which is the direction of planetary rotation, while those farther away (more than 20 million kilometers from Jupiter) revolve in the opposite, so-called "retrograde," direction. This raises a very interesting problem.

SATURN

Famous for its ring system which affords spectacular photographs, Saturn is, like Jupiter, a large, gaseous planet with a fast rotation (10 hours, 15 minutes near the equator; 10 hours, 38 minutes at latitude 36°). Its diameter is 120,000 km (9.4 times that of the Earth). Its mean distance from the Sun, 1,426 million km, varies by 150 million km between aphelion and perihelion, i.e., a distance equal to that separating the Earth from the Sun. The period of revolution of 29 years, 168 days corresponds to a speed of 9.64 km/sec. The high speed rotation causes a flattening of 1/10.

The number of satellites currently known is 17. As for the ring, it is not a single entity; the probes have revealed a multitude of them. As is the case with Jupiter, the satellites and the rings near the planet revolve in the positive direction, that of the planetary rotation, whilst the farthest satellite, Phoebe, 13 million km away from the planet, revolves in the retrograde direction. Furthermore, several satellites circulate along an almost identical orbit, which poses serious problems for the Newtonians, the advocates of stability.

Finally, we should point out that the *Saturn 1* probe found that there were twists in the segments of some rings and formations like the spokes of a wheel. At Saturn's equator, the winds reach a speed of 1770 km an hour, while at latitude 40° north and south the speed is zero; this causes enormous vortices.

URANUS

One of the giant gas planets, with its diameter of 47,140 km (3.7 times the diameter of the Earth), Uranus travels around the Sun at a mean distance of 2,869 million km, with a difference of 265 million km between aphelion and perihelion (i.e., more than the Mars-Sun distance). Its period of revolution is 84 years, 7 days at an average speed of 6.8 km/sec. Its rotation period is 10 hours, 42 minutes according to some and 15 hours according to others. This planet is especially interesting from the point of view of its exceptional position. While the other planets have an axis of rotation nearly perpendicular to the orbital plane, that of Uranus is almost parallel. While the inclination of the equator to the orbit is only 23° in the case of Earth, that of Uranus is 98°. Instead of rotating on its axis in an upright position around the Sun, it is lying down with one of its poles facing the Sun at each solstice.

The planet has five satellites and nine rings. Some revolve in the equatorial plane and hence in a plane almost perpendicular to the planet's orbit, in the retrograde direction—like the planet—while others travel along an inclined plane. The difference between this and the preceding planets is thus appreciable and should be capable of explanation.

NEPTUNE

At a distance of 4,500 million km from the Sun, Neptune, with a diameter of 44,993 km (2.3 Earth diameters) is another of the giant gas planets. Its period of revolution is 164 years, 280 days, at an average speed of 5.4 km/sec. In contrast to Jupiter, Saturn and Uranus, Neptune does not have a regular system of satellites along its equatorial plane. Neptune has only two, the *nearest* of which, Triton (355,000 km away), has a *retrograde* orbit inclined at 160°, while the second known satellite, Nereid, revolves in the *positive* direction although it is *farther* from the planet (5,560,000 km), in an orbit inclined at 28°. This retrograde motion of the nearest satellite while that farthest away revolves in the positive direction is precisely the opposite of the motion of Jupiter's and Saturn's satellites, which calls for an explanation.

PLUTO

The last of the known planets, Pluto is extremely small and its
diameter has not been able to be measured. Situated at a mean
distance of 5,900 million km (7,400 million km at the aphelion and
4,500 million at the perihelion), its sidereal period is estimated at
between 247 and 249 years. In 1978, a satellite was discovered at
about 20,000 km from the planet. It seemed to always present the
same face towards Pluto.

THE COMETS

In addition to the planets and their satellites, the solar system
includes comets consisting of a core surrounded by a fuzzy head.
A tail issuing from the latter expands and lights up as the comet,
coming either from the asteroid zone or from the periphery of the
solar system, approaches the Sun and turns around it before going
off again into the distance. The tail does not follow the comet in
the direction of travel but is driven back in the direction opposite
to the Sun by the "solar wind" or "radiation pressure." The tail
is generally double, one part being formed of dust, the other—much
finer—consisting of ionized molecules swept along by the plasma
and magnetic fields accompanying it. The matter forming the tails
of comets does not join up again with the comet but is dispersed.
Comet tails have been photographed in the process of disintegration. The tail of the 1843 comet measured more than 300 million
km long, or more than twice the distance from the Earth to the Sun.

THE STARS

The stars, including the Sun, are not permanent, immobile bodies.
The majority of those which we see take part in the rotation of the
galaxy, but in addition have their own motions. All of them are actuated by internal movements, irrefutably evidenced by their radiation. All these movements endow them with certain characteristics
which have prompted the astronomers to arrange them into various

classes: pulsating stars (or pulsars), white dwarfs, blue stars, hot stars, red giants, magnetic stars, neutron stars, aberrant stars, flare stars, novae, supernovae. A basic characteristic of all of them is an evolution which is at times very violent. Thus in the case of flare stars, this phenomemon occurs on a far greater scale than solar flares. Novae suddenly become so brilliant that they stand out amongst all the other stars, then they gradually regain their normal luster. Supernovae, which only appear two or three times in a millennium, are even more spectacular phenomena.

Binary stars are extremely numerous. In some regions, one star in two has been identified as a binary. More than 40,000 visual binary stars have been discovered. When the line of sight is favorable, this can assist in eclipses. In addition, large numbers of spectroscopic binary stars have been identified.

Dark nebulas, like the Horsehead Nebula in Orion or the Coalsack in the region of the Southern Cross, attest to the presence of matter in interstellar space.

THE GALAXIES

Galaxies are formations containing up to a hundred billion stars. Thanks to the giant telescopes, superb photographs of these formations in overall motion show them to us head on or in profile in various shapes, the most spectacular of which is the spiral nebula, actuated by a differential rotation which is faster at the center than at the periphery. It is within such a galaxy, the Milky Way, that the solar system is found, in a rather remote area.

The American astronomer Hubble discovered that the spectral lines of these galaxies were all shifting towards the red. He concluded from this that the galaxies were receding at a speed proportional to their distance. This phenomenon has been interpreted as a "flight" of the galaxies and that the universe is expanding. Certain galaxies seem to form clusters.

Matter exists between the galaxies as it does between the stars, in the form of atoms, molecules and dust. Side-view photographs of spiral nebulae, NGC 4565 for example, unquestionably reveal the presence of matter which prevents light from passing through in the equatorial region.

3

Mysteries to Be Solved

THE SOLAR SYSTEM

The solar system is full of unsolved mysteries. Where does the Sun originate from? Where is it going? Why does it rotate on its own axis approximately every 25 days? Why does its diameter appear to shrink? Why does it have planets? Where do they come from? Why do they revolve around the Sun? Where are they going? Do they have a forward or a rear motor force? Whence do they derive the energy which drives them forward and which seems inexhaustible? Why are those nearest the Sun denser than those farther away? Why are they all not at the same distance from the Sun, but distributed as far as Uranus in accordance with a geometric progression with a common ratio of 2, i.e., 1, 2, 4, 8, 16, 32, 64 (Bode's law)? Why are there two exceptions to Bode's law, Mercury being one, and the other being the planets beyond Uranus? Why do some planets rotate on their own axes and others not? Why do some have satellites and others none at all? Why don't they all rotate at the same speed? Why does the Earth rotate in 24 hours and giant Jupiter in less than ten? Why do some satellites revolve in the same direction as the planet while others go round in the opposite direction? Why does Phobos revolve three times faster around Mars (7 hours, 39 minutes) than this planet rotates on its own axis (24 hours, 37 minutes)? Why are comet tails thrown to the side away from the Sun and detached from the comet, while the Earth's atmosphere remains around the planet, despite the latter progressing along its orbit at a rate of 30 km/sec? Why does the Earth have three kinds of years of different durations (the tropical year, sidereal year and so-called anomalistic year)? Why is the Moon progressively receding away from the Earth? Why, instead of revolving on an axis perpendicular to its orbit, does Uranus turn over on its equator and at each solstice turn one of its poles towards the Sun? Why do some planets have rings consisting of debris instead of compact satellites?

Our Earth's past is just as full of mysteries. Why are geological layers radically different from one another? Why are there cyclic climatic phenomena, with glacial periods alternating with tropical ones? Why emergences followed by the immersion of entire continents? Successive formations of mountain ranges? An alternation of the Earth's magnetic fields? An extensive drifting apart of continents formerly united and then broken up? Why, today, is the Earth's crust formed of plates drifting apart or coming together, colliding with each other or riding up over one another? The Earth is a planet whose evolution must accord with that of the solar system.

THE INCOMPREHENSIBLE UNIVERSE

The solar system forms part of a galaxy which is traveling through the universe at a rate of 600 km/sec, a speed that some physicists call "the new ether wind." So what is the motor force causing this overall movement of billions of stars and the clouds between them?

Our galaxy is a spiral nebula whose rotation causes the central elements to recede towards the periphery, as attested by the photographs of spiral nebulae whose arms continually spread out from the core. These formations, including our own galaxy, are expanding. But the same astronomers who point out this galactic expansion and acknowledge that our solar system is part of such a formation assert that, thanks to Newton, the planets dutifully describe closed ellipses around the Sun.

The astronomers found that our galaxy had a "chairlike" form, i.e., that it has an almost flat center in which innumerable stars are found, while those of the periphery move "below" this plane on one side of the galaxy and "above" it on the other side. What could this mean?

If we pass on from the galaxy to the universe, the mysteries are just as innumerable. The one which has provoked, and still provokes, most argument is Hubble's discovery that the spectral lines of the galaxies are characterized by a shift towards the red, which is interpreted as a recession of the galaxies away from one another. We have already referred to the expansion of the universe. But what does this mean? Since the term "universe" includes everything that exists, into what can the universe be expanding?

This concept of expansion has prompted some cosmologists to assume that in the beginning the universe was reduced to an extremely condensed mass which exploded. This explosive creation of the universe was called the "Big Bang," and the "flight" of the galaxies would be nothing more than the consequence of this explosion. But does the universe have a beginning? What happened before the Big Bang and what caused the detonation?

Upon closely studying the cosmologies based on the Big Bang, the famous astronomer Alfven declared that they were no better than the ancient mythologies belonging to the time when Zeus thundered, which caused bitter controversy.

Therefore, in order to clarify our ideas, we must confront the great theories face to face, those of Newton, Einstein and Descartes, and then establish which of them best accounts for the facts without getting caught up in these contradictions.

4

The Newtonian View and the Facts

INABILITY TO EXPLAIN

Faced with all the phenomena characterizing the solar system, Newtonian mechanics proves unable to explain them. It tried to give mathematical expression to the paths of the planets around the Sun and to that of the satellites around the planets, but we shall see further on what to make of this.

Since the most prominent facts are incompatible with Newtonian mechanics, its adherents quite simply ignore them. Thus, Bode's law, with its geometric progression with a common ratio of 2 from Mercury to Uranus, is so mysterious that they are forced to attribute it to chance. Similarly, the displacement of the aphelion, the farthest point of the orbit, is so inexplicable that they have christened the year separating two transits past this point as "anomalistic." Is the Earth really indulging in anomalies instead of respecting Newton's theory? Why does the diameter of the Sun appear to be contracting through the ages, to such a degree that after 200,000 years it would be practically zero? Where do the planets come from, why are they very dense close to the Sun, become gaseous and diminish almost to nothing at the outer limits of the solar system? Why do they have different rotations and why do their satellites behave so differently? No valid reply forthcoming.

And if the stars and planets attract one another in proportion to their mass, why are the galaxies flying apart?

To be sure, Newton's law allows astronomers to carry out purely fictitious calculations on the masses of stars and then to draw fantastic conclusions from them. But is the amusement of astronomers sufficient for a patently baseless hypothesis to become sacred?

THE MYTH OF ATTRACTION

All Newtonian mechanics is based on the hypothesis that "everything occurs as though bodies attracted one another in proportion to their mass and in inverse ratio to the square of their distance." This hypothesis has become a sacrosanct law for a variety of reasons, the first of which is that at a mean distance of 60 earth radii, the Moon "falls" by attraction 60 x 60 = 3600 times less than a body on the Earth's surface. As such a body falls 4.9 m to Earth in the first second, the Moon falls 3600 times less, or 1.36 mm. This "fall" in the course of one second has the effect of making the Moon, traveling forward at 1.02 km/sec, describe an orbit nearly corresponding to its actual orbit.

The law became sacred when, in 1846, the French astronomer Leverrier who had observed some irregularities in the motion of Uranus, attributed them to the interaction of an unknown planet, tried to determine its position, then wrote to Galle, the Berlin astronomer, to ask him whether he could find a planet in the precise region he indicated. Galle actually sighted Neptune and proclaimed the infallibility of Newton's law which had enabled a planet to be discovered by calculation. The whole of Europe went into raptures over the power of mathematics and the virtues of Newton's law.

The law was reconsecrated and became taboo after cosmonauts were sent to the Moon thanks to a meticulous preparation based on Newtonian mechanics. Thus, whoever dares today to doubt Newton's law is considered mentally deficient.

However, it should be pointed out that in the course of the second during which the Moon is "attracted" 1.36 mm by the Earth while advancing along its orbit by 1.02 km, the Earth has traveled 30 km. Newton's calculation only works if the Earth is stopped in its path around the Sun, which clearly exceeds the powers of a man of science, even one as great as Newton. Furthermore, the Sun has traveled 250 km during the same second.

Unfortunately for Leverrier, the data on which he had based his calculations were shown to be false during subsequent observations of Neptune's motion. Leverrier, inspired by Bode's law, had placed the unknown planet at a distance corresponding to the geometric progression with a common ratio of two, or at approximately 6 billion km away, and had assigned it a speed determined by Kepler's law $t^2 = R^3$. But the actual distance is only 4.5 billion km, Bode's law no longer applying beyond Uranus as far as the geometric progression base 2 is concerned, and the speed obviously cannot be that assumed by Leverrier since by very virtue of Kepler's law, it is determined by the distance. So is it just chance that incorrect calculations ended up with a correct result, or is it that Leverrier had discovered Neptune by telescope and wanted to make his discovery more sensational by calculating imaginary interactions and then writing to Galle? It is strange that a Paris astronomer should ask a colleague in Berlin to explore the sky in a very precise direction to locate a planet, as if the Paris observatory only had opera glasses. Caught out at his own game, Leverrier studied Mercury's perturbations and attributed them to the presence of an as yet unknown planet circulating between Mercury and the Sun, a planet which he christened Vulcan and whose transit in front of the Sun he predicted at a precise date. All the astronomers observed the Sun on this date, but no one saw Vulcan, either on this date or any other. Leverrier was utterly mistaken, but this did not prevent astronomers from continuing to believe in the Gospel according to Saint Newton.

As for the lunar expeditions, this achievement—for such it is—has not resolved the question of whether the fall of a body towards a planet's surface is caused by the heavenly body's attraction or by the concentric pressure of the surrounding ether. The effects have been known since Galileo, but not the cause of the falling and acceleration. In his *Principia*, Newton explicitly acknowledged that "Galileo showed that the fall of heavy bodies is in the square of the ratio of the times." It is above all experiments carried out on Earth on the basis of Galileo's data, particularly in aviation, which are decisive. The masses of the heavenly bodies which have been calculated are purely speculative, and when the astronauts went to the Moon, they found that the mass calculations were wrong. But the astronomers did not hesitate in inventing mass concentrations which they ludicrously christened "mascons" in order to adapt reality to Newton's law.

Being prudent in the beginning, Newton asserted that "everything occurs *as if* the heavenly bodies attracted one another in proportion to their mass and in inverse ratio to the square of their distance." Then, impressed by his discovery and unable to say how bodies were able to attract one another, he responded simply: It *is so* since my calculations are correct. Thus a makeshift science was built up without querying whether natural forces of attraction existed and without speculating as to how they were able to act.

Perhaps it would be appropriate to begin by asking ourselves whether there are such things as attractive forces. The General Conference on Weights and Measures (G.C.W.M.) has defined the fundamental standards as follows:

1. The *meter* is a length equal to 1,650,763.73 wavelengths of radiation in a vacuum corresponding to the transition between the 2p10 and 5d5 levels of an atom of Krypton 86 (11th G.C.W.M. 1960).

2. The *second* is the duration of 9,192,631,770 periods of radiation corresponding to the transition between the two hyperfine levels of the ground state of an atom of Cesium 133 (13th G.C.W.M.).

3. *The unit of force is the newton.*
   The newton is the force which imparts an acceleration of one meter per second per second to a body having a mass of one kilogram.

4. *The unit of constraint and pressure is the pascal.* The pascal is the pressure which, acting on a plane surface of one square meter, exerts a total force of one newton on this area.

5. *The unit of work and energy is the joule.*
   The joule is the work done by one newton when the point at which the force is applied is displaced one meter in the direction of the force.

The G.C.W.M. does not define any force of attraction because it knows of none. *The newton is, on the contrary, the force produced by pressure.*

Even the magnet does not possess a force which is essentially attractive since it has two poles producing opposite effects. It manifests the existence of a current which enters at one end and exits from the other. It has been known since Faraday that there is no action at a distance and that in the phenomenon of electromagnetism there exists a field of forces acting together by pressure in a well determined direction.

The physicists who assert that the major part of the mass of atomic nuclei consists of protons which repel one another and neutrons which are incapable of attraction cannot understand how the attraction which they have been taught to believe in holds together the various elements of the nucleus. So they say that the hadrons (protons and neutrons) are *"responsible"* for the cohesion of the nucleus. What could the expression "responsibility," borrowed from the legal fraternity, mean in scientific terms? Instead of acknowledging that there is no such thing as an attractive force, the physicists prefer to resort to obscure terminology.

If, despite its fictions and immense contradictions, Newton's law is still taught today as the great universal law, it is because the squared distance law applying to falling bodies matches reality. But the acceleration of fall does not prove that this falling action is due to an attraction by the purely theoretical center of the Earth. Moreover, the *square* of the distance is inexplicable in the case of attraction, whereas it is easily understood as the effect of concentric pressure of a surrounding medium. A pressure is always inversely proportional to the *surface* on which it is exerted. When directed towards the center of a sphere, it encounters a surface which is smaller and smaller in proportion to the square of the radius ($4 \pi R^2$). Concentric pressure therefore explains the phenomenon of falling bodies much better than attraction at a distance. Moreover, we talk of barometric pressure and not attraction. The deficiency in the value of g, the acceleration constant, which is found on the northern faces of mountains, well demonstrates that the force which causes the fall of a body is the pressure of the surrounding medium which is greater on the sun side than on the side opposite.

In short, if a star or other heavenly body appears to attract bodies in proportion to its mass, it is because this is proportional to the sum of the pressures exerted on its surface by the ambient medium, and if it appears to attract them in inverse ratio to the square of their distance apart, it is because a concentric pressure encounters a surface which is inversely proportional to the square of the radius.

If Galileo, who paid particular attention to the study of these problems, did not conclude, as Newton did, that apples were attracted by the Earth's center, it is probably because he had observed that apples most often fell in a gale and that it was improbable that a gale would increase the Earth's attraction on the apples. Numerous studies of projectiles had taught him that cannonballs did not exit from cannon through an attraction exerted by the enemy, but that it is always necessary to give them a push to set them in motion.

As we shall see later, Einstein did not believe in giant strings of attraction, but he did not investigate why attraction at a distance did not exist. He contented himself with replacing attraction by a geometrization of space, whose explanatory value is no better.

ARE THE HEAVENLY BODIES SELF-PROPELLED?

In asserting that the heavenly bodies attract one another, Newton endowed each with an individual propulsion unit which caused it to travel through space at a fantastic speed and to attract others in proportion to its mass. But from whence do they draw the energy which causes them to advance and to attract? It is easy to say that the Sun attracts the Earth, but when the latter reaches its perihelion, the closest point of its orbit, it rushes away from the Sun at top speed. How can this fantastic attraction all of a sudden cause the Earth to move away? Logically, the law of reciprocal attraction should have the effect of the Earth being increasingly attracted as it approaches the Sun with the result that it should fall into the Sun, just as apples fall to the ground. In order to explain the fact that the Earth, having arrived at the perihelion, recedes away from the Sun which is attracting it, one is forced to assume that there is another force other than attraction. But what is it?

It is here that the Newtonians surpass themselves in evading the real problem. The majority do not refer to it. Others bring in centrifugal force, which is not mentioned in the law. Since this centrifugal force, which does not appear to have existed previously, immediately becomes very troublesome, some physicists deny its existence. Thus in the work *La physique* (*Physics*) published through Dunod by the "Physical science study committee," the following may be read (p. 346): "Centripetal force is real . . . centrifugal force, on the contrary, is not real . . . It is an invention, something which we imagine so that Newton's law of motion may apply in the vehicle's reference system." In order to get out of this difficulty, the authors then speak about the force of inertia. The hypothesis of individual motors propelling the heavenly bodies is all the less admissible in that none has been discovered on Earth, neither the location of the motor, nor its operation, nor where it draws its energy from, nor how it controls direction.

DO THE PLANETS FOLLOW ELLIPSES?

Copernicus realized that the Earth revolved around the Sun and concluded from this that they went round in circles. Then Kepler discovered that the planets were not always at the same distance from the Sun and replaced the circles with ellipses. Why ellipses? This shape is obviously close to reality, but are the orbits closed? His contemporaries were totally convinced of the permanence of the solar system since the Creation was completed and God now rested in contemplation of his immutable work. Newton was also imbued with this preconception of permanence, and all of his mechanics tended to represent universal stability by formulae. But the solar system cannot be regarded as an independent entity. It forms an integral part of an expanding galaxy. One has only to visualize a solar system in the midst of an expanding spiral nebula to realize at once that such a system must necessarily participate in the expansion of the galaxy.

How can it be maintained that the planets describe ellipses when their aphelion, the farthest point of their orbits, moves on in the direction of travel at each revolution? The planet does not describe a closed ellipse, and the anomalistic year of 4 minutes, 45 seconds longer than the sidereal year (revolution of 360°) demonstrates, in fact, that the Earth follows an open spiral.

CAN A PHENOMENON BE INSTANTANEOUS?

Since the mechanism of attraction had never been discovered, Newton drew the conclusion that it was instantaneous. This thesis had even less basis since in 1675 Roemer had determined the speed of light by the observation of Jupiter's satellites and showed that what had been thought of as instantaneous was not so. Common sense requires that every phenomenon must take place in time. To say that a phenomenon is instantaneous is not only to assert that it is unobservable, it is to acknowledge that it does not exist.

Furthermore, it is a contradiction to assert that the phenomenon is instantaneous and then to calculate the effect of this instantaneous phenomenon at the end of a second. Finally, whilst being concerned with calculating what takes place in one second, we should not forget that in cosmology thousands of years are more significant, since it is periods such as these which reveal to us the broad outlines of evolution.

THE SOCIOPOLITICAL AND RELIGIOUS CONTEXT OF NEWTON'S LAW

Newton was deeply religious and had learnedly commented upon the Book of Revelation. He considered Descartes' theory dangerous to religion. In explaining the phenomena of the universe by forms and movements, Descartes had no need of God, except for the initial flick of the finger, for which he was reproached by Pascal. The English intellectuals were then divided into materialists, pantheists and "reformists of a conservative tendency," of which Newton was one. The first used science to challenge the hierarchy and to justify democracy. The pantheists, being of the opinion that God was present in Nature and in Man, considered that priests were no longer necessary. These subversive ideas frightened Newton who, while he acknowledged the reforms introduced by the Anglican faith, wanted to preserve society as it was. In order not to restrict divine intervention to religion and in order to avoid its exclusion from science, Newton thought that God manifested his presence through principles which were the basis of an ordered and harmonious organization of the world. Hence scientific progress became an experimental confirmation of Providence.

Did God intervene directly in attraction? Newton's writings throw some light on this for us. In his long letter Oldenburg on his theory of light and colors which he had presented to the Royal Society in 1675, Newton talks of an "etherial medium" which in addition to a "phlegmatic" part contains various "ethereal spirits" corresponding to electrical and magnetic discharges and to the principle causing gravitation. Elsewhere, he speaks of "the vital aerial spirit." In his letter of 1693 to Richard Bentley, he writes: "That gravitation may be essential and inherent to matter, that one body may act upon another at any distance, across empty space, with no intermediary, is in my opinion such an absurdity that it cannot enter the mind of any man even a little versed in philosophical matters." But it is just such an absurdity, according to Newton, which is acknowledged today by the Newtonians.

Laplace expressly stated that the purpose of his celestial mechanics was to explain all mechanical phenomena in terms of forces acting at a distance. Even after the discovery of radio waves, modern Newtonians have not scrupled to adopt as a creed the very absurdity denounced by Newton and to replace his faith in God or in ethereal spirits by dishonesty.

5

The Einsteinian View and the Facts

ETHER AND EINSTEIN'S THEORY OF SPECIAL RELATIVITY

Let us see whether Einstein's theories are capable of explaining the great mysteries of the solar system and the universe. We have seen that after Michelson's experiment which appeared to mean that 300,000 + 30 = 300,000, Lorentz put forward transformation equations in which he replaced absolute time by local time.

Einstein was greatly impressed by Lorentz's transformation formulae, based on the concept that the Earth was moving through an immobile ether. Hence his main idea was to improve upon these equations. Here is what he wrote in *Ether and the Theory of Relativity*:

> By a marvelous simplification of the theoretical foundations, H. A. Lorentz has succeeded in establishing an agreement between theory and experiment. He achieved this advance in the theory of electricity—the most significant since Maxwell—by divesting ether of its mechanical properties, and matter of its electromagnetic properties . . . As far as the mechanical nature of Lorentz's ether is concerned, it may be said amusingly that the sole mechanical property which Lorentz has left it is immobility. It may be added that the whole of the transformation brought about in the concept of the ether by the theory of special relativity, consisted of it depriving ether of its last mechanical property, i.e., of immobility.

We should point out that Einstein did not divest ether of its immobility in order to attribute movement to it, but to remove from it every mechanical property. In order to make clear what he meant by this removal, Einstein explained that the Maxwell-Lorentz equations are only valid with respect to the sytem of K coordinates in which Lorentz's ether is at rest, whilst in the theory of special relativity, the same equations are valid in the same sense with respect to a quite new system of coordinates, K¹, which is moving in uniform translation with respect to K.

Einstein stated his intention thus:

> The point of view which might be adopted at the outset, would appear to be as follows: The ether does not exist at all . . . Closer reflection, however, teaches us that this denial of ether is not necessarily required by the theory of special relativity. The existence of ether may be assumed, but then it is necessary to forgo attributing a definite movement to it, i.e., one must strip it of the last mechanical property that Lorentz has left to it . . . The principle of special relativity forbids us from regarding the ether as consisting of particles which may be followed through time; but the hypothesis of the ether as such does not contradict the theory of special relativity. It is only necessary to take care not to attribute a state of movement to ether.

Like Hamlet who said "To be or not to be," Einstein declares "To be and not to be." He does confess, however, that such a statement is hard to accept. Consequently he tries to justify it:

> Of course, from the point of view of special relativity, the ether hypothesis seems at first glance to be a barren one . . . On the other hand, one important argument can be adduced in favour of ether: To deny ether's existence means in the last analysis that it must be assumed that space possesses no physical property. But the fundamental facts of mechanics are not in agreement with this concept.

Thus, having said that ether must be deprived of all its mechanical properties, Einstein admits that the fundamental facts of mechanics do not agree with this concept. A more flagrant contradiction one could not imagine.

Einstein continues:

> To sum up, we may say: Under the theory of general relativity, space is endowed with physical properties as such, therefore an ether exists. According to the theory of general relativity a space without ether is inconceivable, since not only the propagation of light would be impossible, but there would even be no possibility for the existence of rulers and clocks, and consequently of spatio-temporal distances in the sense of physics. This ether, however, must not be conceived as being endowed with the property which characterizes the weighable media, i.e., as consisting of particles which may be followed in time. The concept of movement must not be applied to it.

Thus, having asserted that the ether exists, Einstein strips it of all its rank and qualities and dismisses it like a petty criminal. What arrant nonsense this is. How many contradictions there are in this theory which passes for the acme of scientific thought. When Einstein says that without ether the propagation of light would be impossible, he seems to adopt the wave theory of light. But how could this ether transmit the impact causing the light effect other than by movement? Einstein's basic mistake consists of his refusal to allow movement to the ether, since everything in the universe is movement. There is no space without movement, no substance without movement nor movement without substance (as we shall see later on).

Einstein writes: "It would be a notable advance if we were to succeed in uniting the gravitational field and the electromagnetic field in a single representation. Then the whole ether-matter dichotomy would disappear." But there is no dichotomy between ether and matter, the latter being nothing but condensed ether, whose essence is movement. We shall return to this question.

Einstein was made famous by his formula $E = mc^2$ which equated mass with energy (E = energy, m = mass and c = the speed of light). But the concept of energy is derived from that of force and force only exists as movement and because movement encounters resistance.

Some time ago Heraclitus expounded the thesis of the necessity of opposites. Resistance is a force which presses in the opposite direction to another and the whole dynamism is governed by pressure.

Einstein's formula is already present in the works of Galileo, Hooke, Leibniz and Newton in the form $f = ma$, meaning that momentum is equal to the product of mass times acceleration. It is also found in the form of $f = mv^2$, which states that force is equal to the product of the mass times the square of the speed. For $f = mv^2$ Einstein substituted $E = mc^2$. He replaced momentum by energy, which is a matter of vocabulary, and stipulated that the speed was that of light (squared). But why that of light? Einstein confines himself to equations and gives no explanation. However, the profound significance of this formula is understood as soon as it is realized that *the speed of light is none other than that of the internal movement of the transmitting particles*, whether they be particles of ether or of matter. The proof of this is that the speed of propagation of light is of the same order as that of electricity in good conductors (approximately 300,000 km/sec).

Thus, there is no dichotomy between matter and ether, since in both cases the speed of propagation is determined by the speed of the internal movement of the particles. By depriving ether of all movement and all energy (hence also of mass), Einstein set himself in conflict with his famous formula $E = mc^2$, which far from proving the dichotomy between matter and ether, asserts their intimate union. To be sure, Einstein only applies his formula to material masses and disregards interstellar and intergalactic substance, whose mass, taken in its entirety, is much greater than that of the visible masses. Thus he only takes account of a minute fraction of reality and ignores the true motive force of the universe.

CONTRADICTIONS IN THE THEORY OF SPECIAL RELATIVITY

It is easy to see that Einstein contradicts himself when, having written that without ether there could be no propagation of light, and thus referring to the wave theory of light, he constructs his thesis upon the emission of photons. All the arguments that he laboriously develops regarding the train, the carriages, the traveler moving along in the train and the observer on the embankment, in order to demonstrate that people do not see phenomena in the same way, the whole of his theorem on the addition of velocities, all his formulae are based upon the theory of photon emission.

Now, in proceeding along its orbit at a rate of 30 km/sec, the Earth produces no more light than a motor car traveling along with its headlights off. Hence there are not two lights to be added together since the Earth does not produce any, and Michelson's experiment is based on a badly stated problem. Even if the Earth produced light like a car driver putting his headlights on while traveling, there would be no advantage in adding speeds since it is not the speed of the vehicle which produces the light, but the specific impact, and once the impact has occurred, the vehicle always races behind the wave which is propagated at a rate of 300,000 km/sec. The speed of the vehicle running behind the wave should not therefore be taken into consideration in the wave theory of light. Moreover, De Sitter has demonstrated that the speed of transmission of the wave is not affected by the movement of the object *producing* the wave. On the other hand, it may be influenced subsidiarily by the overall movement of all the particles participating in the *transmission* of the wave, as Fizeau revealed in his experiments on the speed of light in fluids in motion. But this is quite another problem.

In short, *interferences ought not to occur in Michelson's experiment* for two reasons. The first is the wave nature of light, and the second is that the ether, far from being immobile, as Michelson assumed, actually drives the Earth along its orbit, just as it drives the galaxies and causes them to revolve. Accordingly there was no difference in speed between the surrounding ether and the Earth proceeding along its orbit. *The negative result of Michelson's experiment is proof of the accuracy of the wave theory of light and the driving of the Earth by the surrounding ether.* The existence of an immobile ether would imply that of a fixed center of the universe, since things can only be immobile with respect to a fixed point.

It is only proponents of the theory of corpuscular emission who are surprised by the negative result of Michelson's experiment. It is only they who have to add 300,000 + 30 and are astounded at only finding a total of 300,000. Neither the photoelectric effect nor other phenomena can invalidate the correctness of the wave theory. The emission theory is based on the error of regarding a displacement of basic light particles through space as being a real displacement, whereas the phenomenon of light is always due to an undulation of the ether produced by a specific impact.

In laying down as a principle that the speed of light is a constant, Einstein draws the conclusion that it is impossible to determine absolute simultaneity. Such a conclusion, implying that it is impossible to measure phenomena, is the negation of measuring science. But this then dumbfounds the physicist who has passed his whole life with rulers and clocks for measuring the world and putting it into formulae. There is a flagrant contradiction between the assertion of the constancy of the speed of light and the denial of simultaneity. Bergson, in fact, took a stand against this denial of simultaneity, but since he did not show why the result of Michelson's experiment should be negative, the argument was an empty one.

It is intriguing to note that having laid down the constancy of the speed of light as a fundamental principle in his theory of special relativity, Einstein was forced to acknowledge that his fundamental principle was not valid in gravitational fields. Accordingly, very embarrassed, he writes this:

It might be thought that this consequence overturns the theory of special relativity and with it the theory of relativity in general. But, in actual fact, this is not so. It can only be concluded that the theory of special relativity may not lay claim to a limitless sphere of validity; its effects are only valid insofar as the influence of gravitational fields on phenomena (on light for example) may be disregarded.

Thus, having admitted that the fundamental principle on which the theory of special relativity rests is only partially correct, Einstein effortlessly proves that in relativity, partially correct principles are just as good as any others.

THE ETHER AND EINSTEIN'S THEORY OF GRAVITATION

The theory of special relativity being only a kinematic representation, Einstein tried to include gravitation in his theory of general relativity. But how did Einstein conceive of gravitation? Here is what he said of it (*Relativity*, chapter 19, "Gravitation"):

> When the question is asked: Why does a stone which has been thrown in the air, after rising up, fall to the ground?—the following reply is usually received: Because it is attracted by the Earth. Modern physics gives the reply a little differently for the following reason. Through a more accurate study of electromagnetic phenomena the concept has been reached that *direct action at a distance does not exist*. If a magnet attracts a piece of iron, we should not be content with the concept that the magnet acts directly upon the iron across the empty space separating them, but rather, following Faraday, we must imagine that the magnet continually produces *something physically real* in the space surrounding it which is designated by the term "magnetic field." This magnetic field in turn acts on the piece of iron in such a way that the latter tends to move towards the magnet. We do not wish to discuss here the justification for this intermediary notion which in itself is arbitrary. Let us merely observe that thanks to it we can give a more satisfactory theoretical explanation of electromagnetic phenomena than without it, particularly with regard to the propagation of electromagnetic waves. The effects of gravitation are to be conceived in a similar fashion. The action of the Earth upon the stone takes place indirectly. The Earth generates a gravitational field in its vicinity, which acts upon the stone and causes its falling movement.

How can one pile so many contradictions one on top of the other? To write that the magnet produces something physically real—a magnetic field—and then a few lines farther down to write that this is an arbitrary notion? Is it reality which is arbitrary or Einstein's theory? How can the latter refer to Faraday, the discoverer of induction, acknowledge the existence of electromagnetic waves and dispute that ether possesses movement? And lastly what is the meaning of: "The action of the Earth upon the stone takes place indirectly. The Earth generates a gravitational field in its vicinity, which acts upon the stone and causes its falling movement''? Is not such a sentence indicative of Einstein's confusion when he is, in fact, disputing Newton's attractive force, whilst still having need of it, acknowledges the existence of electromagnetic waves yet refuses to give a name to the undulating medium?

There are so many contradictions in the theory of relativity that an exhaustive examination would form the subject of endless commentaries. Let us simply point out that in his work on relativity, in chapter 30, entitled "Cosmological difficulties in Newton's theory," Einstein asserts that the law on attraction does not correspond to the facts, particularly with regard to the distribution of the stars in the heavens, for if the law were true, there ought to be a center of the universe where the density of stars would be at a maximum. But Einstein does not examine why Newton's law is fundamentally false. He restricts himself to making some slight alterations to it.

His thesis postulating that the photons emitted by the stars are deflected from their trajectory by the curvature of the universe in the vicinity of the Sun implies, in fact, that these photons are attracted by the Sun's mass, but is expressed differently by a curvature of the region surrounding the Sun. Whatever its expression, the idea that the photons draw nearer to the Sun when they pass close by is in flagrant contradiction with the undeniable phenomenon of the "solar wind" which drives comet tails away from the Sun.

As for this new geometrization of space with its curvature close to stars, this is in flagrant contradiction with the images of the galaxies supplied by the great telescopes. Of course space is not flat but consists of a quantity of galaxies and stars much more dense at the center than at the periphery and actuated by a rotational movement terminating in tremendous expulsions of mass.

It is unnecessary go into all the details of Einstein's theory or into the serious contradiction between the thesis of special relativity and that of general relativity with regard to the constancy of the speed of light. Suffice it to limit ourselves to the essence of the matter which is the following statement: Basing himself on Minkowski's thesis of four dimensional space-time, Einstein does away with time as a reality and replaces real time t with the imaginary magnitude $\sqrt{-1}ct$ which is proportional to it. His conclusion is as follows (*La relativité*, p. 172, Edition Petite bibliothèque Payot): "It seems more natural to represent physical reality as a four-dimensional being instead of representing it, as has been done until now, as the evolvement of a three-dimensional being."

The more one rereads this sentence, the more one wonders how a man of science could write such nonsense. In this way, by getting rid of evolvement, Einstein hopes to represent physical reality and to build a science of the future. Whilst more than two millennia ago Heraclitus proclaimed that "nothing is, everything is becoming" and Plato wrote that "knowledge is a probable opinion of becoming," in the midst of the twentieth century Einstein is constructing a theory which excludes "becoming" or evolvement. One wonders whether science exists to sanctify the meaningless phrase and how without evolving, four-dimensional Einstein could have died. Could it be that Time, which does seem to be getting a little old, did not get to know of Einstein's theory or did this physicist of genius suddenly have a flash of common sense which caused him to pass away?

The idea of a space-time continuum is not absurd in itself. Quite the opposite, substance and movement are intimately connected, as Hooke very aptly observed nearly three centuries ago, but this space-time continuum, which is none other than the intimate union of substance and movement, is, in point of fact, evolvement. If the time shown by clocks is only the relationship of its movement with another movement taken as a standard, this standard is nothing but evolvement, since it is movement.

Einstein's theory has been successful for the same reasons as Newton's. It appears to provide a mathematical solution to badly formulated problems. Overturning conceptions based on solid work establishing the wave nature of light, its revolutionary character could not but appeal to all those who think that all change is progress and who wish to sail with the balloon of fashion, especially if the theory contains a lot of hot air. Finally, it has the double advantage of allowing the most abstruse commentaries whilst excluding from their comprehension all sensible persons who believe that the real world should be capable of being explained by the study of real phenomena. Through this exclusion of common sense, physics accordingly becomes a field reserved for the super-intelligent. Little does it matter that the abolition of ether may be in contradiction to the indisputable existence of electromagnetic waves. Little does it matter also that the constancy of the speed of light—a fundamental principle of special relativity—is not valid in the theory of general relativity. From the moment that incoherence is laid down as the fundamental principle of science, any rash follies are permissible.

Of course, Einstein was not always wrong. He knew that one could utilize nuclear energy to make an atomic bomb. Having passed on the process to a team capable of manufacturing the device and having discovered the dreadful result, he launched solemn appeals to the States to implore them in the name of morality not to use it. But why was Einstein so worried about the future evolvement of mankind when according to him evolvement did not exist? His qualifications in the sphere of destruction do not mean that the alleged confirmations of the theory of relativity are other than mere assertions. Thus, at the time of an eclipse of the Sun, the positions of the stars which appear close to the darkened solar disc are different from the positions calculated by the astronomers. Einstein deduced from this that the photons emitted by these stars when close to the Sun were influenced by a curvature due to the Sun's gravitational field, and he very earnestly calculated this curvature. But this presupposes an actual displacement of the photons which is incompatible with the wave theory of light. The solar wind, whose effect on comets is beyond question, on the contrary, causes particles close to the Sun to recede away from it. Einstein's theory is therefore in contradiction to the facts. The difference in position is explained by the wave theory of light and not by the photon emission theory. The speed of light transmission depends on the one hand upon the speed of the internal movement of ether particles, and on the other hand upon the overall movement of these particles within the solar vortex. As Fizeau has shown, the transmission speed of the light wave varies according to whether the particles which are *transmitting* the wave phenomenon are moving in the direction of the transmission of the initial impact or in the reverse direction. Einstein's theory cannot be confirmed by hypotheses based on the actual displacement of photons, whereas the wave theory alone truly accounts for the facts.

Various theoreticians have attempted to explain what Einsteinian curvature ought to mean. They have then represented the world by a plane surface, hence a two-dimensional world, and not one with four dimensions, as Einstein maintained, or three-dimensional as required by reality, and they have placed spheres on this surface, which by their weight cause subsidence of the areas surrounding them. They then released some smaller balls whose paths were obviously deflected by these depressions. Such is the physical representation of the curvature of the Einsteinian universe. But observation teaches us that the universe is full of galaxies and stars which, instead of causing depressions in the areas surrounding them, emit vast quantities of matter in the form of various types of radiation, in particular as the solar wind or radiation pressure.

Claims have also been made of seeing confirmation of the theory of relativity in the anomalies in Mercury's motion. We shall see later on that the explanation is quite different.

INADEQUACY OF NEWTON'S AND EINSTEIN'S THEORIES

Hubble's law, according to which galactic centers recede from the observer at speeds proportional to their distance, undermines the basis of both Newton's law and Einstein's theory. Newton's law was based on the assumption of a stable universe and tried to explain why the planets followed ellipses around the Sun. Hubble's law demonstrates that all the large formations of the universe are unstable. The photographs of spiral nebulae reveal that these formations are actuated by a rotational movement which causes both the billions of stars of a galaxy as well as the interstellar "clouds" to move progressively away from the center towards the periphery. These galaxies are therefore expanding, and this proves the evolution of the whole universe—stars and interstellar space—i.e., its *evolvement* that Einstein claimed to abolish.

In order to explain how the theory of relativity, which excluded evolvement, could be applied to an expanding universe, Einstein was content to merely add a new component to his equations. Here is what he says in his *"Réflexions sur l'électrodynamique, l'éther, la géométrie et la relativité"* ("Reflections on electrodynamics, ether, geometry and relativity") (Gauthier-Villard, p. 99):

> The objection which may be made to this solution (regarding the mean density being equal throughout the four-dimensional universe) is that one is obliged to introduce a negative pressure, for which there is no physical justification. In order to make this solution acceptable, I have first introduced a new component into the equation in place of the pressure that I have just referred to, which from the point of view of the theory of relativity is admissible.

Einstein then adds a constant A to his gravitation equations, which he calls the universal or cosmological constant.

Einstein's astuteness is one of the same order as that of the cloth merchant who made a fortune by selling his material at the same price per meter as he bought it. In order to make a profit, he just had to lengthen his meter when he bought and shorten it when he sold. He thus sold more meters than he had bought. It was easy. Someone was bound to think of it.

In short, Einstein prefers to add a new element to an equation in order to represent a factor which has no physical justification, rather than revise his manifestly false conceptions regarding the structure and evolution of the universe.

It is understandable that caught up in such contradictions, Einstein should have passed the last thirty years of his life trying in vain to construct a general dynamics just as applicable to atoms as to the heavenly bodies, strong interactions, weak interactions, electromagnetic interactions and to gravitation. He kept on running into inconsistencies, just as Empedocles of Agrigentum did in the past, juggling with his four fundamental elements—earth, air, fire and water.

In short, both Newtonian gravitation and Einstein's theory contain contradictions which prevent them from giving a true picture of the world and from answering the most important questions that can be put. Where does the Sun come from? Where is it going? Why does it rotate on its own axis? Why does its diameter seem to be shrinking? Why is it surrounded by planets and they by satellites? Where do these heavenly bodies come from? Why are the planets distributed from Mercury to Uranus in accordance with a geometrical progression with a common ratio of 2? Why do satellites near the planet revolve in the same direction as the planet's rotation, and the satellites farthest away, in the reverse direction? Why does the Earth's atmosphere remain around the planet while it races along at 30 km/sec? Why is it not driven towards the side away from the Sun, like comet tails, etc.? Phobos, which revolves around Mars three times faster than the planet on its own axis, shows the inanity of these theories.

Of course, these theories give great delight to the astronomers by making it possible for them to set out figures and Greek letters in fantastic and complicated formulae and to create a pseudo-scientific jargon by speaking of relativistic electrons (why not "collectivistic," "Salvationistic" or "Marxist-Leninistic" ones, as though the particles had learned to conform to Einstein's theories?). They have even imagined gravitational waves which must retreat in order to attract. Obviously they cannot explain the mechanism of these retreating waves, but they venture to invent gravitational collapse. A house subsides or collapses because its foundations are too weak. Celestial bodies collapse in on themselves because they attract themselves too much and end up by forming a black hole, which some astronomers have started to look for in the heavens in vain. Thus, by dint of trying to harness the universe beneath the iron yoke of permanence whilst it is, in fact, in perpetual evolution, modern science has ended up in a tunnel, from which who knows when it will escape, so persistent are its preconceived opinions.

6

The Cartesian View and the Facts

THE EXISTENCE OF VORTICES CONFIRMED

Starting from the idea that in a continuous universe, where everything is in motion, the various particles must inevitably collide with one another, Descartes concluded that they must form huge, contiguous vortices. Since then, the giant telescopes constructed during the last century have allowed innumerable galaxies to be photographed, often occurring in the form of spiral nebulae. Descartes' brilliant inspiration was thus confirmed, whilst no one was ever able to photograph the attractive force of Newton, whom the textbooks can only show in portrait.

To be sure, the photographs generally only show the image of the centers of the galaxies, and there are many astronomers who assert that the galaxies are separated by empty space. But space is an empty word, whose meaning varies according to the needs of argument. On the other hand, ether well and truly exists since light and radio waves disclose its undulation. As for its properties, we shall examine these later on. Despite the scant interest shown by men of science in the vortices, which nevertheless represent the major units of the universe, their study is highly instructive.

FIRST FINDING: THE HIERARCHY OF THE VORTICES

The solar system is a vortex forming an integral part of the galactic vortex. Similarly, the planets with their retinue of satellites form

51

vortices within the solar system. Thus there are not only juxtaposed vortices, as Descartes has delineated them, but a hierarchy of vortices, the smaller ones being incorporated in the larger.

SECOND FINDING: CENTRIFUGAL FORCE

The study of vortices in constant expansion, corroborated by examination of the nebulae whose arms are moving progressively away from the center, reveals to us that there is a centrifugal force in the whole vortex, with this particular characteristic that in the central region it is proportional to the radius. This centrifugal force accordingly reduces the force being exerted in a circular direction, which reduces the speed of the planets as they get farther away from the Sun (Kepler's law).

THIRD FINDING: REVERSAL OF FLOW AT THE PERIPHERY

In tending towards the outside, the centrifugal force which encounters a resistance causes a reversal of flow starting from a critical distance from the center. Normally developed vortices thus have a movement at the periphery which is in the reverse direction to that of the center. But where does the force come from which causes the vortices to turn?

QUANTA AND THE DYNAMICS OF ETHER

Modern physics is presently founded upon two basic constants, the Einsteinian (inconstant) constant of the speed of light and Planck's constant. These two constants are generally accepted as being the two immovable pillars on which the whole scientific edifice rests. Here is what Planck himself said of them (*"Initiation à la physique"* ["Introduction to Physics"], Flammarion, p. 162):

> Two great new ideas contribute to setting the tone in the new physics: these are on the one hand the theory of relativity, and on the other hand the quantum theory. Each has contributed its share to the fruitful overthrow of ideas, *but they none the less*

*remain total strangers to one another and up to a point are even
opposed to each other.*

These two immovable pillars are therefore not as immovable as some
physicists would like to pretend.

The discovery of Planck's constant has, in fact, a much greater
importance than the theory of relativity, which is basically no more
than a dispute over reference systems. Max Planck had the merit
of pursuing his experiments on radiation to the point of discovering
that the infinitely small was finitely small, or, as Planck himself explains, the fragmentation of energy cannot be pursued beyond a certain
limit. The discrete value to which a manifestation of energy corresponds he called a quantum of action and succeeded in calculating
this extremely small value very accurately. He also found by experiment what the ancients had conceived theoretically, namely that a
finite quantity cannot be divided into an infinite number of portions.
It is this concept which had prompted the ancients to adopt the idea
of indivisible atoms.

Planck was aware of the importance of quanta. In fact he wrote
the following (*op. cit.*, p. 165): "The quantum hypothesis does not
confine itself to contradicting generally accepted old ideas . . . it
contradicts the most fundamental postulates of classical physics.
Hence it is not just a simple modification as in the case of relativity,
but a real revolution in ideas." He acknowledges, however, "that
there are fields in physics, such as the vast field of interferences,
where classical theory is proven by the most accurate measurements
down to the smallest details and where quantum theory, at least in
its present form, falls down completely."

Why does quantum theory fall down with regard to interferences?
Could it be that quanta do not exist? On the contrary, Planck is
so certain of their existence that he writes as follows (*op. cit.*, p.
48): "Although the inmost nature of these dynamic quanta still remains rather enigmatic, in the light of the facts known at present,
it becomes difficult to doubt that they exist in some way, *since what
can be measured must necessarily exist.*" In which case, since they
exist and since interferences also exist, why does the theory break down?
The reason is as follows: Planck unfortunately drew the conclusion

from his discovery that the universe was discontinuous and that light was transmitted by corpuscular emission. Planck's error lay in not investigating the concepts of force and substance more thoroughly. He wrote as follows (*op. cit.*, p. 96): "The concept of force has proved itself pre-eminently useful in formulating the law of motion, but in itself it has not advanced science in the least."

How can Planck write this while calling his quanta "*dynamic*" (p. 48)? There is the same contradiction with regard to substance: "The concept of substance, though it has played an important role since time immemorial, has nevertheless not always contributed to progress" (p. 152). However, he writes a little farther on: "Quanta behave like actual atoms of substance in their impact with matter."

Formal physics makes a very clear distinction between particles and force fields, the latter (electromagnetic or gravitational) being regarded as continuous and devoid of particles. But there can no more be force fields without actual force units than there can be carrot fields without carrots. It is, in fact, in the concept of "field" that the great misunderstanding lies. It is time to put an end to this unfortunate distinction between field and particles and the misconception that a field can be formed of nothing and that particles imply discontinuity. The distinction disappears as soon as it is recognized that a particle is not an inert unit, but that it is dynamic and behaves according to the laws of dynamics.

A particle is dynamic because it is actuated by an internal movement which pushes it to extend until it meets a resistance, i.e., a force equal to its pressure. Actuated by an internal movement, it is thus a unit, but the existence of dynamic units does not imply discontinuity, since these units crowd against one another and interpenetrate, exchanging movements, as we shall see later.

Quantum theory is not therefore incompatible with the wave conception of light since it does not mean that the universe is discontinuous. Quite the contrary, in attesting the existence of dynamic particles, it reveals to us what it is that undulates in the wave theory of light and removes any distinction between ether and matter. In a word, quanta prove that force fields, generally regarded as being immaterial, are formed of real, dynamic particles. Moreover, Planck was on the right road when he wrote (*op. cit.*, p. 12): "The former distinction between matter and ether seems to be becoming blurred. Electromagnetism and dynamism are far from being irrevocably opposed . . . There are good indications that the fields in question will end by amalgamating into a single general dynamics."

The General Conference on Weights and Measures has abandoned any distinction between ether and matter by adopting definitions of standards of space, time, pressure and work. It defines real space and real time, not Einstein's space-time. It does not recognize any attractive force. The newton is, on the contrary, a force produced by pressure.[1]

The choice of a wavelength as a standard of space does away with the distinction which formerly obtained between the material, regarded as the only existing reality, and the immaterial, devoid of an existence of its own. Since what can be measured must necessarily exist (Planck), henceforth the immaterial must exist by virtue of the same reason as the material. Planck's constant (h) may therefore be regarded as the true standard of energy and mass of the universe, i.e., $h = mc^2$, and $m = \dfrac{h}{c^2}$ applies both to interstellar space as well as to material bodies.

THE UNIVERSAL SUBSTANCE

In serving as the universal standard, immaterial space, defined both by its extent and by its property of undulating, implies the intimate union of substance and movement. This union is such that there is no such thing in the universe as substance without movement nor movement without substance.

Hence, one might perhaps be forgiven for thinking that there is only one universal substance, whose essence is movement and all of whose properties are derived from movement. Assuredly, the concepts of matter and substance are not identical since the physicists have restricted the concept of matter to certain bodies formed from well-defined particles. The much wider term of substance as defined

1. The definitions appear on page 32.

by Spinoza and used by Planck does not admit of any essential difference between material bodies and intergalactic space since both have no essence other than movement. Just as the solid, liquid or gaseous state depends upon the conditions in which particles are found, so these conditions determine the properties of ether and its ability to change into matter as a result of the concentric pressure exerted by the ambient medium. Thus, matter is, in fact, nothing more than a condensed form of ether which is capable of reassuming its non-condensed form.

Various experiments have shown that a pair of electrons—one negative, the other positive—are capable of dematerializing under radiation. Conversely, radiation may be transformed into a positron-negatron pair, which literally constitutes a materialization of radiation.

The discovery of spin, the internal movement of all particles, by Uhlenbeck and Goudsmit, is one of the most important of this century, since it enables us to understand the inmost nature of particles, which are not inert globules, but units actuated by an internal movement. The magnitude of the spin vector is measured in $\hbar$ units, the symbol being read as "bar h," equal to Planck's constant divided by 2π. The photon, which is not a stray particle but the quantum of radiation, has a spin equal to one. Planck's action quantum is actually the unit of movement serving as a standard for the whole universal substance.

GENERAL DYNAMICS AND UNIVERSAL EVOLUTION

The essence of substance, the movement which exists within every particle and pushes it to expand until it encounters a resistance equal to its own pressure, turns the particle into a force, so that the universe is dynamic as a whole and in each of its parts.

Pressing against one another, the particles, which are actuated by an internal movement, tend to equalize their movements, like two cogwheels turning in opposite directions, the one with the greatest movement yielding up its excess to that having the least. It is in this contact and this exchange of movement, substance and mass (since there is no movement without substance) that the continuum of the universe consists. The continuum is therefore not a simple juxtaposition of inert units. It is this existence of real units of force possessing their own movement and mass which attests to Planck's constant. All of space, whether material or immaterial, is a force field composed of real force units.

By choosing not a rigid body but a phenomenon as the universal standard, the G.C.W.M. has substituted movement for inertia, evolvement for stagnation. But how does evolvement come about?

Compressed by its neighbors, every force ought to take the form of a sphere due to the equalization of movements. But inside the particle, movement, which cannot be spherical, consists of a series of superposed rotational movements around an axis. The antagonism between the plastic resultant due to the equalization of external pressures and the dynamic resultant of the internal movements leads to an *evolution of the particle* which flattens at the poles and expands at the equator, as demonstrated by any experiment on the rotation of a mass. *Continuum, dynamism and evolution are therefore the essential and indissociable characteristics of the universe.* Moreover, is not movement synonymous with change?

The evolution of particles by expansion is spectacularly illustrated by radioactivity. But this phenomenon has still to be interpreted correctly. In the previously cited work, Planck wrote this:

> Here we have an atom of uranium which has remained absolutely passive and invariable in the middle of atoms of the same type surrounding it through countless millions of years; suddenly, without any outside cause, in an interval of time whose brevity defies all measurement, this atom violently explodes . . . and the phenomenon is accompanied by the emission of an electromagnetic radiation whose fineness exceeds that of the most penetrating Roentgen rays. Next to it, the neighboring atom which is identical to it in every way, remains unaffected by the cataclysm; perhaps it will have to wait millions of years before undergoing the same fate. Add to this the fact that all the attempts made to influence this phenomenon, such as raising or lowering the temperature, have failed completely, and we find that there is not

the least hope of obtaining an idea of the laws regulating atomic
disintegration phenomena. Yet the theory of this disintegration
is one of the most important in physics (*op. cit.*, p. 53).

It is obvious that there is not the least hope of understanding
the laws regulating atomic disintegration if it is assumd "a priori"
that the atom has remained passive and unchanging for millions of
years, whereas knowledge of the forces which act within particles
reveals to us that these particles evolve and that radioactivity is a
manifestation of this universal evolution. Elsewhere we have described the process, which may be summarized as follows: atoms
are assemblages of particles actuated by an internal rotational movement (spin). Although they do not all have the same density or the
same volume, nevertheless they all behave like larger or smaller
cogwheels. When in contact along their equators, they revolve in opposite directions and form a couple; when in contact at their poles,
they turn in the same direction and form an axis. Atoms are
assemblages of couples and axes. The internal movement in time produces a flattening of the particle at the poles and an expansion at
the equator. In the case of bodies formed from a limited number
of particles, this expansion is not distinguished by any spectacular
phenomenon, but nonetheless does involve a slow and continuous
expansion of the atom. In the case of very complex bodies, the conflicts of expansion compel certain particles to alter position within
the atom. When they reach a position of 45° with respect to their
axis (or their plane), their movements are no longer in harmony since,
in order for this to be so, at their equator the particles must revolve
in opposite directions while at the poles they must revolve in the
same direction. When the particles reach halfway between the equator
and the poles, the discordance in their movements causes their expulsion. *Such is the cause of radioactivity. It lies in the conflicts
of particle expansion within the atom.*

Thus, radioactivity is not due to a sudden change of mood in
certain atoms, but to the dynamics of particles taking part in universal evolution. In addition, it means that atoms of the same element
are not all at the same stage of evolution. Radioactivity is nothing
but a spectacular phenomenon displaying the evolution of all the

particles of the universe whether they be condensed into atoms or form interstellar and intergalactic space. All of them evolve because their essence is movement. It is his failure to understand this universal evolution within particles which led Planck to believe in the stability of atoms throughout the ages.

Moreover, it should not be forgotten that the Earth is not surrounded by empty space, but is compressed by the ambient ether. It is the concentric pressure of this medium which keeps together the various particles constituting the atom. It is not the hadrons (protons and neutrons) which are responsible for the cohesion of the atomic nucleus, it is the pressure of the medium which compresses the Earth, makes it turn on its own axis and revolve around the Sun at the speed of 30 km/sec. Finally, we should point out that recent experiments have revealed that high pressure slowed down the process of radioactivity, which means that it is indeed evolution in the form of expansion which is the cause of radioactivity.

It is also evolution which is the cause of the different properties of protons and electrons which, contrary to the assertions of the physicists due to misinterpreted experiments, have the same mass, i.e., the same quantity of movement, since protons and electrons are in contact at their equators and are ceaselessly exchanging their quantity of movement like two cogwheels. Situated on the periphery, the electron is subject to less pressure from the ambient medium than the proton which is located at the center of the atom. Now, the expansion of particles through time is proportional to their volume, which increases the diameter of electrons and reduces that of protons since an increase in the diameter of the largest wheel automatically causes an increase in the speed of the smallest and therefore reduces its diameter. This is why the electron's density diminishes with time, while that of the proton increases, although their quantity of movement remains equal, but differently distributed amongst the particles.

GALACTIC EVOLUTION

After Hubble discovered a shift of the spectral lines of the galaxies towards the red, which he estimated as proportional to their distance from the observer, there was talk of the "flight" of the galaxies and from this discovery the hastily drawn conclusion was that the universe was expanding. But such a conclusion is unacceptable since it implies the existence of a second universe into which the galaxies were expanding. Based on this "flight," the theory of the "Big Bang," the primeval explosion of the whole universal mass, once concentrated into an extremely small volume, is no more acceptable since it presupposes the absurd idea of a beginning of the universe starting from a non-time, a zero time, and the creation of movement from immobility. The indisputable phenomenon of evolution excludes any beginning.

On the other hand, if by being defined as encompassing everything that exists, the universe cannot be expanding, innumerable photographs show us an undeniable galactic expansion. Thus, by their rotational movement, with their arms which increasingly recede from their center, the spiral nebulae disseminate incredible quantities of movement and mass in a variety of forms. Sometimes these are huge satellites, like the enormous luminous mass expelled by M51 of the Hunting Dogs (*Canes Venatici* constellation). Generally, they take the form of countless stars progressively receding from the galactic center. These expulsions of solid or gaseous matter from the core occur at speeds of several thousands of kilometers a second. They are accompanied by particle dispersion in the form of cosmic radiation. Finally, the transfer of movement and mass takes place even more subtly from particle to particle of the ambient medium, thereby increasing the movement and mass of particles situated in the most remote regions. Equatorial expansion and progressive flattening are substantiated by profile-view photographs of the nebulae, which also confirm the expulsion of substance from the center by the dark bars seen at the equator. *Hence, within the spiral nebulae an evolution occurs which is characterized by a new distribution of masses at the expense of the central zone and to the gain of the most distant regions.*

But how far does this displacement of mass extend? The galaxies being numberless and the universe continuous, the expansion of some clashes with that of others, so that there are different galactic zones: first of all the center, generally visible, very condensed, the explosion of which continually disperses a large proportion of mass into the much more extensive, and for the most part invisible, median zone, which receives this mass and partially transmits it with some delay towards the periphery, the confrontation zone, where these excess masses accumulate and collide with one another. The confrontation zone is composed of vast eddies in which the particles of diverse galaxies are reciprocally compressed. In spite of the extreme speed of transmission, the wave flux which sets out from the center towards the periphery lasts some thousands of years (our galaxy's radius is generally estimated at more than 15,000 parsecs, i.e., more than 50,000 light-years), the equalization of movements between particles therefore never results in immobility nor in a uniform distribution of masses, but *produces a universal family of eddies characterized by the dispersal of old galaxies in favor of new ones which will be dispersed in their turn.* Thus the galaxies are not all of the same age and were not all created on the day of the "Big Bang." They are clusters which form, then disappear, to be replaced by others.

While the term "age" can be applied to galaxies since it refers to clusters which form and break up, it cannot be appropriate to the universe as this would then mean that there was a time when it did not exist and it would have originated from nothing.

When light hits a galactic zone which is contracting, it does not pass through, as witnessed by the dark bars along the equator of galaxies viewed in profile. There are then dark regions in the universe, but they never result from "black holes" which are products of the minds of those who support accretion and relativity. A hole is, by definition, an empty space and not an accumulation of matter, which is always due to pressures. Everyone is aware that a house collapses because its foundations are too weak. In the gravitational collapse of the Newtonians, masses collapse in on themselves because they are so dense that they can only attract by collapsing inwards! However, if there are masses in the universe which collide and concentrate, it is not because they attract one another, but because they are compressed by galactic expansion.

The significance of the red shift in the spectral lines of the galaxies can only be understood if one rids oneself of the erroneous ideas in general currency regarding the structure of the universe. Since ether manifests its presence through its undulation, it is absurd to deny it any attribute of movement or any mechanical property, as Einstein did. It is no more acceptable to regard it as immobile, as Michelson assumed when he conducted his experiment. Such an assumption would imply the existence of a fixed center of the universe which until now no one has discovered. Granted that the universe is homogeneous, since it is formed of a single substance, all of whose particles have the same essence—movement—this is far from producing the same effects when the conditions are not the same. While there are spiral nebulae, there are clusters which have a different movement and a different form. If, strictly, one may speak of homogeneity, with many reservations, it cannot be a matter of regarding the universe as isotropic, i.e., having the same properties in all directions. Photographs of spiral nebulae which are very condensed at their centers, clearly show that this is not the case. In the solar system, NASA has established a pressure of 95 atmospheres on Venus; while on our planet, the properties of a solid mass are clearly different from those of liquids and of the atmosphere. Furthermore, a spiral nebula viewed in profile appears like a sphere flattened by its rotation. Now, in a sphere subject to the concentric pressure of the ambient medium, the particles near the center are more compressed than those on the periphery. Interstellar and intergalactic space therefore have particles which are very different as regards their properties, according to their position in the formation of which they are a part.

Finally, let us not forget that all radiation is a transfer of energy. Bathers who get a touch of sunstroke know this as well as researchers who have to use solar energy.

In order to gain an idea of what actually does happen, let us follow the path of a wave emitted by a galactic center. The ether which oscillates in transmitting this energy does not bring to a given point of arrival the total energy emitted at the point of departure. A portion has been dissipated in the surrounding space. The most obvious proof of this dissipation is the diminution in luminosity with distance. The ambient medium retards the movement of particles taking part in the displacement of the wave. Such a braking action necessarily modifies the wave length, just like the swell which forms when the waves of a storm come into collision with the rest of the ocean. Why should ether waves not undergo the swell effect which reduces the height of the wave and increases the length?

To be sure, the comparison is not perfect, since in a light wave, the phenomenon is generally continuous and the frequency remains generally the same at the emitting source. Nevertheless, the fact remains that the transmission of a wave motion cannot occur without friction with the ambient medium.

Light is only propagated in a straight line for a short distance and in an isotropic medium. Refraction and diffraction attest to the influence of the transmitting medium. Not only is the wave retarded by the particles of the ambient medium, but it is also affected by the overall movement of these particles. When the latter are animated by a vortical motion, as is the case in the spiral nebulae, the wave necessarily takes part in this motion and describes a curve. This curvature of the light wave due to the vortical movement of the transmitting ether has nothing to do with that which Einstein devised since the cause is quite different. The displacement with respect to a straight line is so obvious that at the time of solar eclipses, the stars which appear close to the Sun's disk while it is obscured by the Moon are not found in the position calculated by the astronomers. In following a curve instead of a straight line, the wave travels a longer path, and when the light source is a galaxy, the lengthening of route due to vortical movement is much greater than within the solar vortex. It is even greater when the wave has to cross several galaxies. In addition, the red shift is larger.

Evolution within a galactic vortex is not restricted to rotation. The ejection of movement and mass from the center towards the periphery increases the mass of the particles of the invisible median zone, which is, in fact, the zone which transmits the light effect, so that the latter is affected on the one hand by the increase in volume of the particles transmitting the wave, and on the other hand, by the expansion of this median zone due to the increase in particle volume, with the result that the overall rotational movement lengthens the curve.

In short, numerous factors intervene in the propagation of the phenomenon of light and produce a shift of spectral lines towards the red. This shift is thus not due to a "flight" of the galaxies in an empty, "ad hoc" created universe, but results from the structure of the universe, which is continuous, dynamic and evolving through the dissolution of old galaxies in favor of new ones formed from the debris of the old which break up in their turn, for, as Heraclitus pointed out, "nothing is, everything is becoming."

This shift is thrown into sharp relief by certain objects called quasars (quasi stars) which transmit radio waves one hundred million times more intensely than our galaxy. These extremely intense sources of flow of very high energy particles are extragalactic, as shown by the exceptional shift of their spectra towards the red (up to z = 3.53 whereas the record shift reached by a normal galaxy is 0.9). The astronomers consider that, allowing for their distance, the quasars are the brightest objects in the universe, their luminosity being capable of reaching a hundred thousand times that of our galaxy. They are the source of an enormous release of energy.

Some galaxy clusters are quite remarkable. Thus sixteen out of the few hundred known clusters are X-ray emitters as powerful as a hundred billion suns (10^{44} ergs emitted per second) and present an insoluble problem for the Newtonians: the visible mass of the cluster on average represents only a few percent of the mass required to maintain the cluster by gravitational attraction. Hence they speculate on where the missing mass is to be found.

For a long time it was believed that our galaxy was much quieter than others where real catastrophes had been witnessed, including radiogalaxies, Seyfert galaxies, quasars, jet emission (M 87, 3 C 371), exploding centers (M 82), etc. But these catastrophes are not restricted to the other galaxies. Thanks to the advances in non-optical astronomy, we are forced to recognize that our galaxy is the site of violent events: high-speed displacements of matter detected by radiotelescopes, infrared and intense X-ray emissions, etc.

STELLAR EVOLUTION

Within a galaxy which is expanding and breaking up, the stars can but take part in the same sort of evolution. Just as the galaxies are not all of the same age, so the stars are distinguishable from each other by their degree of evolution, determined to a large extent by their position in relation to the galactic center and by the particular conditions in which they find themselves. Also, the observer is struck by their diversity. The astronomers distinguish between white dwarfs, blue stars, red and yellow giants, visual or spectroscopic double stars, numerous eclipsing binaries (several tens of thousands have been identified), periodic, pulsating, irregular, unstable, neutron, X ray and explosive variables. All of them are characterized, however, by one common phenomenon—their radiation—proof of their evolution, which nevertheless may be very different. While some appear to radiate in a constant manner, there are many which manifest themselves in abrupt changes in brightness or radio emissions characteristic of violent events, particularly in the case of novae and supernovae.

Novae are stars whose explosion is manifested in a considerable increase in brightness and radio emissions. Some are observed every year and to date several hundred are known. Supernovae are giant novae, which are much rarer (not even one a century).

In September 1972, Cygnus X-3's brightness suddenly increased one thousandfold then rapidly diminished to reappear 15 days later before becoming scarcely perceptible. In 1975, the star Nova Herculis 1934 revealed itself to be an eclipsing binary. Radio pulsars, X-ray pulsars and X-ray "jump" stars reveal the violence of the phenomena whose source these stars are.

In the case of SS 433, two jets thrown out by the star in two opposite directions indicate by their different spectral lines that the body is rotating, with a probable period of 164 days, while the speed of ejection of the jets would be incredibly high (approximately 80,000 km/sec). This expulsion of jets from a star is a particularly instructive phenomenon, since it reveals to us that reality is exactly the opposite of attraction.

The considerable number of double and eclipsing stars proves that these stars evolve like the Sun, as we shall see.

THE EVOLUTION OF THE SOLAR SYSTEM

THE SUN

Located in a rather peripheral zone of an evolving galaxy, called the Milky Way, the solar system is drawn along by the latter's gyratory movement at a speed of 250 km/sec which applies to both the ambient ether and the stars bathing in this ether. With its planets, it itself forms a vortex of ether.

At its center, the Sun is formed of the same substance as the rest of the universe in a concentrated form. Since the essence of this substance is movement, the Sun takes part in the motion of the galaxy characterized by the expansion of everything pertaining to it. Stars containing a large quantity of substance necessarily have a considerable expansion since the concentric pressure which created them has engendered extremely dense elements. Expansion then produces a phenomenal disintegration of these elemnts in the form of radioactivity. It is this radioactivity which, by causing impacts on the ambient medium, is the origin of solar radiation.[1] The light which reaches us from the Sun and the stars is the most vivid proof of the expansion and disintegration of the Sun. Similarly, what the astronomers call the "solar wind" is essentially composed of protons and electrons, originating from this disintegration.

The planets and satellites together with all the bodies composing them do not escape the general phenomenon of expansion, but the densest elements having generally exploded in the Sun, radioactivity is much less in their case. Although less spectacular, the expansion of non-radioactive bodies is nonetheless widespread and continuous. The moving ether which causes the Sun to travel at a rate of 250 km/sec equally causes it to rotate on its own axis in approximately 25 days. It is this which also draws the planets around the Sun and causes them to rotate on their axes.

As for the Sun's origin, one has only to consider the movement of the spiral nebulae to see that all the stars are progressively receding from the center as a result of centrifugal force. The Sun, which belongs to such a galaxy, therefore originates from the galactic center and in the course of time is receding away from it.

---

1. The fusion of hydrogen atoms into helium atoms which occurs at the Sun's surface and which some physicists believe to be the fundamental phenomenon of solar evolution is a secondary phenomenon arising from the radioactivity of the most complex bodies and from the enormous heat resulting from it. It is well known that this tremendous heat is essential to the fusion of light elements into heavy elements.

Whereas the Newtonian conception of a discontinuous universe in which the void is traveled by heavenly bodies which are self-propelled, driven by front or rear motors, strings of attraction or an initial flick of the finger, does not explain these fantastic movements nor other mysteries of the solar system, they become perfectly intelligible as soon as it is accepted that the universe is continuous, that the Sun is drawn along by the galactic vortex and the planets by the solar vortex. Therefore, let us examine these mysteries more closely.

BODE'S LAW REGARDING THE PLANETARY DISTANCES FROM THE SUN

The orderly distribution of the planets in the heavens is a phenomenon of the highest importance. The German astronomer Bode recognized that they were not scattered at random but that from Mercury to Uranus their distances from the Sun were in accordance with a geometrical progression with a common ratio of 2, i.e., 0 for Mercury, 1 for Venus, 2 for Earth, 4 for Mars, 8 for the asteroids, 16 for Jupiter, 32 for Saturn and 64 for Uranus. The supporters of permanence and Newtonian mechanics are completely unable to explain this progression and attribute it to chance, which is not an explanation. However, it is clear that this distance, which doubles each time from one planet to the next, is due to a very definite phenomenon.

If one rids oneself of the preconception of the permanence of the solar system and studies its evolution, if, in addition, it is known that within a vortex there is a centrifugal force proportional to the radius, it will be realized that during a given time, the planet which is located at distance 1 moves away from 1 to 2, that which is located at 2 moves away from 2 to 4, that at 4 moves away from 4 to 8, etc. In short, far from being due to chance, as is believed by the advocates of the permanence of the solar system, *Bode's law is the major law revealing its evolution by expansion.*

One has only to go back in time to realize that Bode's law means that the Sun, which, let us not forget, is rotating, periodically expels a planet, so that the nearest to it are the youngest, and the farthest away are the oldest, contrary to what is taught by "official" science, dominated by the theory of a condensing nebula creating the solar system.

Within the solar system, Bode's law is the homologue of Hubble's law on the recession of the galaxies. The base 2 geometrical progression is met again in the central structure of spiral nebulae, which confirms that the centrifugal force proportional to the radius exists in all vortices. It is also found in the distance of the satellites with respect to their planets, though a little less exact, due to the fact that the planet's speed of rotation increases when it becomes gaseous (see figure 7).

The base 2 geometrical progression of Bode's law contains two exceptions: one at the beginning in regard to Mercury, the other at the end in regard to the planets situated beyond Uranus whose recession within a given time is constant instead of being progressive.

The need to add a constant to Mercury means that, at the time of expulsion, the planet cleared a very great distance in one bound, as indicated by the very eccentric orbit of this planet which is 48 million km from the Sun at its perihelion and 70 million km at its aphelion. The effect of ejection is also apparent in the case of Phobos, a recently expelled satellite of Mars, as we shall see later on. The constant must be added to all the other planets, which indicates that they have all been ejected in the same way.

For practical reasons, in order to obtain the figure 10 for the relative Earth-Sun distance, astronomers multiply the geometric progression by three. This measure in no way alters the fundamental characteristic of Bode's law, which is its geometrical progression by a common ratio of 2. The constant added to the tripled progression is 4. Thus the relative distance is obtained, which has only to be multiplied by 14.8 million km (i.e., 1/10 of the actual Earth-Sun distance, whose relative distance is 10) in order to obtain the exact distance of the planets.

Planetary Distances from the Sun According to Bode's Law

Planets	Geometric progression C.R. = 2	Tripled	+ Constant 4	Relative distance	Distance in millions of km	
					According to Bode	As calculated by astronomers
Mercury	0 x 3 =	0	+ 4 =	4 x 14.8 =	59	58
Venus	1	3		7	104	108
Earth	2	6		10	148	149
Mars	4	12		16	237	227
Asteroids	8	24		28	414	410
Jupiter	16	48		52	770	777
Saturn	32	96		100	1480	1426
Uranus	64	192		196	2900	2869
Neptune	128	384		388	5742	4395
Pluto	256	768		772	11426	5898

From Mercury to Uranus, the agreement between the results of Bode's law and the distances calculated directly by the astronomers is all the more striking in that the latter are averages, the planets not being always at the same distance from the Sun, which varies by a few million kilometers between aphelion and perihelion.

INTERPRETATION OF BODE'S LAW BY THE NEWTONIANS

In *La nouvelle astronomie, science de l'univers* (*The New Astronomy, Science of the Universe*), published in 1977 by twenty-two specialists, under the direction of a member of the Paris Academy of Sciences, Bode's law is looked upon as nothing more than "an empirical law, to be regarded merely as a simple mnemonic means of recalling the distances of the planets from the Sun" (p. 105).

When a motorist drives through one red light, then a second, third, fourth, fifth, sixth and seventh red light, any sensible person thinks that either he does not know the meaning of red lights or that he is thumbing his nose at the world. It would not enter anyone's head to assert that these are mnemonic means intended to enable the motorist to calculate distances traveled. To say that Bode's law, with its base 2 geometrical progression which repeats itself seven times, is a simple, mnemonic means of recalling the distances of the planets is to believe that Nature has taken pity on tired astronomers. When then are these proponents of a new astronomy going to realize that so-called universal attraction does not exist and that Newton's law is rubbish? When will they grasp that the solar system is not static but is evolving prodigiously in accordance with the expansion of our galaxy? That the Earth is receding progressively away from the Sun, and that neither Newton's hypotheses nor Einstein's theories will prevent it from so doing?

In order to show just how absurd modern theories are becoming, one contributor to this *New Astronomy, Science of the Universe* writes the following in a book entitled *"The Exploration of the Universe* ("L'exploration de l'univers") (Hachette, p. 10):

> If it is to be assumed that the universe is inflating like a balloon, it should still be realized that the atom does not undergo expansion no more than the standard meter or the Earth-Sun distance. It is only the distances between galaxies or between clusters of galaxies that increase during the expansion of the universe; for it is only on the scale of these distances that the notion of a homogeneous universe retains any sense. *The galaxies themselves maintain fixed dimensions*, just as coins covering the balloon before it had been inflated would not change.

How can anyone maintain that the galaxies, once compressed into an initial lump, whose diameter then at a minimum and extending today over thousands of light years, preserve fixed dimensions within an expanding universe? It would seem that a minimum of common sense ought to prompt this astrophysicist to conclude in favor of galactic expansion and expansion of the solar system incorporated in such a galaxy.

Bode's law is, in cosmology, what is called in geometry a "pons asinorum." Whoever does not succeed in crossing it will never understand anything of the science he is attempting to learn.

EXCEPTION TO THE BASE 2 GEOMETRICAL PROGRESSION WITH REGARD
TO THE PLANETS SITUATED BEYOND URANUS

As can be seen from the previous table, the distances cease to
double beyond Uranus and increase from one to the next by 1500
million km both between Uranus and Neptune, and Neptune and
Pluto, which is the same as the distance between Saturn and Uranus.

This exception to the geometrical progression has the effect of
discrediting Bode's law in the eyes of the official astronomers for
whom mathematics are paramount. But mathematics do not rule the
universe. They exist to measure it and, if possible, to explain it. If
one begins from the principle that the universe must conform to
human mathematics, one denies oneself any possibility of explaining phenomena. Since the exception to the geometrical progression
exists, it is not a matter of disputing the existence and the value of
the progression which is repeated seven times, but of determining
the cause of the exception.

Just as the constant value 4 representing the distance from Mercury to the Sun revealed the existence of a particular event—
expulsion—which affects all the planets, since 4 must be added to
all of them in order to ascertain their relative distance, so the constant recession of 1500 million km during the same lapse of time
for the farthest planets is attributable to a cause which must be
studied closely.

CHANGE IN DIRECTION OF THE VORTEX AND URANUS' SOMERSAULT

The second exception to Bode's law is explained by the change
in direction of the solar vortex which revolves in the positive direction up to Uranus and begins to turn in the reverse direction from
this planet onwards, the centrifugal force having driven the ether
particles towards the outside to such a degree that the movement
is reversed. From the Earth onwards, the vortex movement may still
seem positive though slower, but the reversal is made manifest by
the position of the planet whose axis of rotation is almost parallel
to the ecliptic, instead of being nearly perpendicular as in the case
of the other planets. The change in direction of the current causes

Uranus to tip over and its North pole becomes its South pole and vice versa. Here again Newtonian mechanics, which excludes ether, is unable to explain this abnormal position of Uranus, which becomes perfectly intelligible as soon as it is asked how a planet which has revolved on its own axis and around the Sun in the positive direction for millennia behaves when the vortex movement changes direction. It must inevitably tip over. Such is the natural sequence of evolution.

By changing direction, the centrifugal force ceases to be proportional to the radius and becomes constant, which corresponds to Hubble's law, according to which expansion is almost the same in every direction. Only the vortex centers have a special régime due to the conditions of centrifugal force.

The change in direction which takes place within a vortex is also apparent in the case of Jupiter where the satellites nearest to the planet revolve in the same direction as the latter (positive direction), while the others revolve in the retrograde direction. The same applies to the satellites of Saturn. Those of Neptune attest to a double reversal: first of all that of the solar vortex, since the planet revolves in the retrograde direction whereas the planets closer in revolve in the positive direction; then the reversal of the planetary vortex, since Triton, which is close to Neptune, revolves in the retrograde direction like the planet, while Nereid, which is farther away, revolves in the positive direction. The motions of Neptune's satellites are therefore exactly the reverse of those of Jupiter or Saturn, whose near satellites revolve in the positive direction and whose farthest ones revolve in the retrograde direction. These reversals, which are inexplicable in Newtonian mechanics, clearly demonstrate that the movement of the heavenly bodies is due neither to chance nor to their having self-propulsion, but is the result of the movement of the ambient medium—ether—whose vortical movement alters direction starting from a critical distance from the center.

THE MYSTERY OF PHOBOS

The mystery of Phobos is one of the most intriguing of all those needing to be clarified. This satellite of Mars, which is only 6000 km from the planet, corresponding to the diameter of Mars or half the Earth's diameter, and which revolves around Mars in 7 hours, 39 minutes, or three times faster than Mars' own rotation (24 hours, 37 minutes), baffles all Newton's and Einstein's calculations. It is obviously difficult to determine whether it is Mars' motor which drives Phobos, or Phobos' motor which drives Mars. Any explanation outside of evolution is obviously impossible. If, on the contrary, we refer to the latter and allow that rotation causes the periodic expulsion of planets by the Sun and satellites by the planets, we realize that this phenomenon imparts a tremendous speed to the ejected mass. Phobos was expelled like a bullet from a gun.

It is interesting to put together the speed of Phobos and the special characteristics of Mercury, which follows a very elongated trajectory. It is also quite interesting to compare the photographs of Phobos and Mercury which are characterized by numerous impacts of masses which have produced craters. Mercury, Mars, Phobos and the Moon all have tortured surfaces, attesting to the fact that expulsion of planets by the Sun and satellites by the planets is a phenomenon of extraordinary violence whose repercussions appear to be just as great upon the expelled mass as upon the expelling mass.

The photographs taken of Phobos by NASA reveal that this satellite is covered in striae and grooves, indicating undeniable abrasion which can only have taken place with Mars. This means that Phobos was once in contact with Mars and then moved away from the planet. These striae are a proof of its expulsion.

In the nineteenth century, Mars was depicted as covered in canals. Recent photographs show long fractures more or less filled with lava. On the Earth's surface, the majority of geologists accept the existence of plates which move with respect to one another. It is probable that if we could remove the water from the surface of our globe, it would appear fractured and would bear some resemblance to that of Mars, although the Moon's expulsion is well prior to that of Phobos and the water has been greatly instrumental in leveling the fractures.

The existence of water on Mars in the past, even though at present everything is frozen, means that Mars was closer to the Sun than it is today. This is an additional proof of the progressive recession of the planets.

Neither is it without interest to discover that the planet Mars, which rotates on its own axis at almost the same speed as the Earth, has a magnetic field one hundred times less than predicted. It is not illogical to assume that when a planet becomes deformed due to its rotation, the dipole existing within it splits into a quadripole and that the opposition of these two dipoles contributes to the expulsion of the satellite in the form of a discharge. Mars' magnetic field would thus have been discharged at the time of the expulsion of Phobos. In support of this thesis it may be noted that the inclination of Mars' dipole is 73° with respect to its axis of rotation. In this connection we should observe that the Earth's magnetic field is very variable, that its declination has altered appreciably over the last few centuries and that its intensity is decreasing. Would this indicate that a quadripole is gradually taking the place of the dipole?

CAUSE AND EFFECTS OF THE ROTATION OF THE PLANETS

It is not by chance that the planets rotate on their own axes. But the Newtonians are again extremely reticent about this phenomenon. It is therefore essential to examine the cause and effects of rotation.

As Bode's law indicates, the planets are moving progressively away from the Sun from which they originate. At the time of their expulsion, they are in a very condensed ambient medium which grips them as though in a vise. The probes sent to Venus have revealed a pressure of 95 atmospheres. Thus compressed, the planets revolve around the Sun always keeping the same face towards it, just as the Moon does around the Earth.

As they move away, the vise slackens its hold, and as speeds in the solar vortex diminish with increasing distance from the center (Kepler's law: $t^2 = R^3$), the planet is driven along more rapidly on its diurnal face closest to the Sun than on its nocturnal face and so begins to rotate on its own axis. In order to visualize what happens, just place a pencil between the palms of both hands and rub them together. The differential movement causes the pencil to rotate.

The greater the diameter of the planetary vortex, the more significant is the difference in speed of the solar vortex, which explains the much faster rotation of the giant planets by comparison with the smaller ones. This difference in speed manifests itself even when the Earth is at its aphelion. Farther from the Sun than at perihelion, the less compressed vortex expands and revolves faster.

Driven along faster on the diurnal side than on the nocturnal side, the planet first begins to rotate in the retrograde direction, i.e., in the reverse direction from the solar vortex which drives it. In so doing, the planet itself forms a vortex which in turn drives along a part of the fluid mass which surrounds it. With increasing expansion, this fluid mass, revolving in the retrograde direction, then forces the planet to rotate in the positive direction, sinee a normally formed vortex is composed of two zones each revolving in opposite directions.

This explains why Venus, which is just getting under way, is rotating very slowly in the retrograde direction whereas the Earth and the planets farther away are rotating in the positive direction, like the Sun, while from Uranus onwards the planets again rotate in the retrograde direction, the solar vortex having altered direction.

When, under the pressure of the ambient medium which compresses it and drives it along, the center of the fluid vortex consists of a dense mass rotating at the same angular speed, the medium which imparts its rotation to it is retarded by the mass that it is driving, so that in such a vortex we have the following zones starting from the center:

1. The dense central zone rotating as a unit more slowly than the surrounding medium which drives it along.
2. A fluid zone retarded by the dense mass and animated by a movement which accelerates increasingly towards the exterior, as the central braking action becomes less and less felt.
3. The fluid zone having the fastest vortical movement.
4. A fluid zone whose movement slows down as the distance from the center increases, in accordance with Kepler's formula, $t^2 = R^3$.
5. The peripheral fluid zone with movement in the reverse direction from the central rotation.

The fact that it is the ambient medium which determines the rotation of the planets is attested by the existence of the equatorial winds rotating in the stratosphere much faster than the planets. This phenomenon is widespread. As for the difference in pressure within a vortex, this is brought out in a particularly striking manner by the outer ε ring of Uranus which describes a very elongated orbit. Its width varies linearly with its distance from the center of the planet. It is 20 km when it is near the center and 85 km when it is farthest away. Furthermore, its major axis is moving through 1.4 degrees per day, which indicates a recent expulsion.

Although the magnetopause is much closer to the planet on the diurnal than on the nocturnal side, it must not be far from representing the limits of the planetary vortex. As soon as the limits of the planetary magnetopauses are ascertained with sufficient accuracy, as well as the differences in particle speeds between the diurnal and nocturnal sides, it will undoubtedly be possible to establish mathematically the cause and effect relationship between this difference in speed and the planetary rotation.

The rotating planet becomes deformed, flattens at the poles, expands at the equator and in its turn ultimately expels a satellite, in fact it does so periodically since the rotation continues. Hence the number of satellites that the planets have, increases the farther they are from the Sun. This then is the actual mechanism to which the planets and satellites owe their origin.

The photographs reproduced in *La Terre s'en va* (*Earth's Flight Beyond*) well illustrate the various stages of the progressive deformation of a rotating spherical mass up to its splitting.

It is not beyond the bounds of possibility that the photographs of twisted coils and spoked wheel formations taken by *Voyager 1* around Saturn represent an ejection of matter from the planet, just as the spectra of SS 433 allow us to conclude that such an expulsion has occurred from a star.

EVOLUTION OF THE PLANETS

Just as the spiral nebulae are much denser at the center than at the periphery—as the photographs show—so a planet moving away from the Sun passes into a more and more tenuous medium. Being less compressed, the planet expands and becomes gaseous. This is the case with gaseous Jupiter, which rotates at a different speed at the poles from that at the equator. The same applies to the planets farther out.

In expanding, the planetary vortex is subject to a greater speed differential on its diurnal face than on its nocturnal face (Kepler's law previously referred to), which explains why Jupiter and the more remote giant planets rotate on their own axes in about 10 hours even though they have far greater volumes than the Earth whose rotation takes 24 hours. It is for this reason also that the Earth, after reaching aphelion, the farthest point of its orbit, is a little less compressed by the ambient medium and that with the expansion of the planetary vortex, the speed of the Earth's rotation is slightly faster than at the perihelion where the planetary vortex is denser.

The evolution of the planetary vortex is well displayed in the characteristics of Jupiter, Saturn, Uranus and Neptune which, with their satellites, represent four successive stages of this evolution.

While Jupiter's nearest satellites, like the planet, rotate in the positive direction along a plane which is slightly inclined with respect to the ecliptic, the peripheral satellites which circulate in the retrograde direction are located on extremely inclined planes. The tipping movement of the vortex therefore starts on the outside.

With Saturn—the second stage—we discover that the rings and satellites close to the planet again rotate in the positive direction, like the latter, whereas the farthest away has a retrograde motion; but the tipping movement of the vortex is clearly more advanced, for Saturn's axis of rotation is much more inclined than that of Jupiter (26°44' instead of 3°05').

In the third stage, Uranus is in the process of turning a somersault, passing squarely from the positive to the retrograde direction. Its equatorial plane is inclined at 98° with respect to its orbital plane. The same applies to its satellites. Thus it is the whole vortex, planet included, which is in the process of tipping over.

In the case of Neptune, the somersault is completed. It has a slight obliquity. Its rotation is retrograde, the same as that of Triton, the nearest satellite, while Nereid, the farthest, makes its revolution in the positive direction, attesting once more to the fact that at their periphery the vortices rotate in the opposite direction to that of the central region.

It is interesting to compare this evolution of planetary systems to the "chair" shape of our galaxy, which also seems to be in the process of turning over, implying that it belongs to a larger vortical system.

A planet which periodically ejects a satellite loses more and more mass. Passing into a less and less dense medium as it moves away from the Sun, its vortex becomes less and less coherent. Its outer satellites are caught up by the solar vortex, in which they are projected in the form of comets, the resultant between their former movement around the planet and that of the solar vortex causing them to describe a very elongated orbit. The planet is finally reduced to a core which gradually disintegrates (see Neptune and Pluto). Such is the fate of planets, which pass through a very dense phase in their youth (Mercury, Venus, the Earth, Mars), then through a phase of expansion (Jupiter, Saturn, Uranus), followed by rapid decay (Neptune and Pluto).

Can the Progressive Recession of the Planets and Satellites Be Observed?

Due to the motion of all the stars, and in particular of our own point of observation—the Earth—calculations of distances and their changes are not easy. It is, however, certain that with modern methods, repeated and correctly interpreted observations will soon enable us to determine exactly to what extent the planets are receding from the Sun, and the satellites from the planets. We already have some significant elements at our disposal. The first element: the Sun's diameter appears to be shrinking with time. The second element: the point of the Earth's orbit farthest from the Sun (aphelion) is shifting forwards each year in the direction of the Earth's travel, with the result that we have three kinds of years, all of different durations: the tropical year, which serves as the basis of our calendar, about twenty minutes shorter than the sidereal year, which marks a complete revolution, and the "anomalistic" year, so-called by the astronomers who have not grasped its significance. Representing the time between two aphelia, this is 4 minutes and 45 seconds longer than the sidereal year. Finally, we have the third element: it is established that the Moon is progressively moving away from the Earth. Let us examine these three elements a little closer. Aside from this we should note that in *La Terre s'en va* (*Earth's Flight Beyond*) we adopted as the basis for calculating the Earth's age various physical and geological phenomena, such as the diminution in the speed of light ascertained by Michelson, and the formation of coal layers.

THE SUN'S DIAMETER APPEARS TO BE SHRINKING

The astronomers of the Greenwich Observatory who have been checking the diameter of the Sun since 1836 have established a contraction of 0.1 percent in a century. An identical finding results from measurements made since 1846 by the Washington Naval Observatory. If this tendency should continue at the same rate, the diameter of the Sun would be practically nil at the end of 200,000 years. Since they consider such a conclusion absurd, the astronomers presume that this tendency reverses itself at the end of a certain time. But their thesis is founded solely upon their desire to avoid an absurd conclusion. In fact, they are acting like Ptolemy who invented epicycles in order to explain the movement of the planets because he was blind to the motion of his own observation point, the Earth. Our modern astronomers commit the same error in believing that the Earth-Sun distance remains the same from one year to the next, whereas it is increasing. The contraction of the Sun is not a reality, it is an illusion due to the progressive recession of the Earth with respect to the Sun. Everyone knows that an object appears smaller as it moves away.

THE THREE KINDS OF YEARS

The year which serves as the basis of our calendar does not correspond, as is commonly believed, to a complete revolution of the Earth around the Sun. It corresponds to the return of the seasons, i.e., to the passage of the Earth to what the astronomers call the "vernal point" characterized also by the equality of day and night.

This is the tropical year, slightly shorter by about twenty minutes than the sidereal year which corresponds to a complete revolution marked by an ideal line between the Sun and a star which is considered to be fixed. The sidereal year is merely equal to a complete orbit of 360°. The third kind of year that the astronomers have called anomalistic is marked by the return of the Earth to the farthest point of its orbit from the Sun. It is 4 minutes and 45 seconds longer than the sidereal year.

The three kinds of years of differing durations must be explained other than by just giving them different names. The Earth's axis of rotation, which is not exactly perpendicular to the plane of its orbit, is gradually straightening up a little each year. Our planet thus passes the vernal point a little earlier each year. Hence the difference between the tropical year on which our calendar is based and a complete revolution around the Sun. As for the anomalistic year, characterized by the shifting of the aphelion in the direction of travel, this is due to the annual enlargement of the Earth's orbit under the influence of centrifugal force. In order to illustrate this behavior of the planet, we have only to look at what happens when a car leaves the road on a sharp bend precisely because of centrifugal force. It is diverted *forwards* in the direction of travel and *to the outside* along the resultant of its own speed and the centrifugal force. Thus the displacement of the aphelion (the farthest point of the orbit from the Sun) in the direction of travel is explained by the enlargement of the Earth's orbit. Once more it is found that the progress of the Earth is determined by the forces of the medium drawing it along.

THE MOON IS MOVING PROGRESSIVELY AWAY FROM THE EARTH

The study of the Moon's eclipses dates from earliest antiquity. It is said that Thales of Miletus learned to predict them from the Babylonians. Now, after almost three thousand years that eclipses have been recorded, comparisons indicate that the Moon was closer to the Earth than at present and revolved around it faster. The official explanations consist in assuming either a transfer of energy from the Earth to the Moon, or an effect of the tides caused by the Moon on the Earth. In the case of the transfer of energy, this would have had the effect of causing the Moon to pass over to a more distant orbit, which is scarcely compatible with attraction. The tides would, on the contrary, cause a slowing down of the Earth's rotation and therefore a diminution in the force of attraction exerted on the Moon.

These explanations do not take account of universal evolution in the form of an expansion which causes a continuous recession of the planets away from the Sun and of the satellites away from the planets. Such is the true cause of the Moon's recession.

We should observe in connection with the tides that if they were due to the attraction of the oceans by the Moon and by the Sun, they would demonstrate that pseudo-attraction does not act instantaneously since high water does not correspond to the passage of the Moon and tides are almost non-existent in the Mediterranean and in the middle of the oceans, just where there is the greatest mass. The tides are due essentially to the Earth's rotation, which is not the same for a liquid as for a solid, the least cohesion of which causes a certain delay. This does not exclude the Moon's influence, since the pressure of the ambient medium upon the Earth's sphere is modified according to whether this pressure is exerted on a medium formed solely of ether or whether a solid body, such as the Moon, comes between. Similarly, the radiation pressure which impinges on all the planets and drives comet tails away from the Sun cannot be the same if it hits the Earth head-on or if it first hits the Moon, which in some instances, when situated between the Earth and the Sun, acts as a shield. The tides nonetheless remain a phenomenon due essentially to the Earth's rotation which takes place from West to East, and to geographical peculiarities (depth and distribution of shorelines) which condition the ebb and flow of the oceans. The major currents, such as the Gulf Stream, and temperature differences may further accentuate or diminish the phenomenon. The whole dynamism of the tides is governed by a combination of pressures. There is not a scrap of attraction.

THE SHATTERING OF PLANETS AND SATELLITES

In studying Bode's law, some astronomers found that there was a "hole" in the base 2 geometrical progression between Mars and

Jupiter and began to scan the sky to find the missing planet. But instead of discovering a new planet like the others, they found a quantity of debris which they called "asteroids."

It seems then that instead of having been expelled in one piece, the planet has shattered into numerous fragments. The presence of the asteroids at the location indicated by the base 2 geometrical progression between Mars and Jupiter confirms the thesis that this progression is a reality due to a definite fact, the expansion of the solar system with the consequent periodic expulsion of a planet by the Sun. It invalidates the most widely held thesis that the solar system was formed by the condensation of a nebula. It makes plausible the idea that at the time of the expulsion, which is in a way an explosion, the expelled mass may split into fragments.

The same phenomenon occurs during the expulsion of satellites by planets. The most famous example of this dispersion of debris is undoubtedly Saturn's rings, of which there are remarkable photographs. Saturn is not the only planet to be surrounded by rings. They have been discovered recently around Uranus and more recently still around Jupiter. Moreover, these planets do not just have a single ring. There are several relatively close to one another and closer to the planet than the compact satellites.

This peculiarity may be explained thus: as the planet moves away from the Sun, the concentric pressure diminishes, as can be seen from the photographs of spiral nebulae. The cohesion of the planet inevitably diminishes. Thus from Jupiter onwards they appear to be essentially gaseous, except perhaps for the invisible central part. This lack of cohesion has the effect that the expulsion of satellites, which takes place in one piece when the planet is denser, only occurs in the form of countless bits of debris.

In view of this reduction in cohesiveness, which seems to be especially noticeable beyond Mars, it may be wondered whether the asteroids were fragmented debris at the time of their expulsion from the Sun, or whether a planet which was more or less compact up to that point started to explode during the expulsion of a satellite.

The impacts which are found on Mercury, Mars, the Moon, Phobos and on some of Jupiter's satellites clearly confirm that an expulsion consists of a tremendous explosion with all its consequences.

THE EARTH'S ATMOSPHERE

Why does the Earth retain its atmosphere despite advancing along its orbit at a rate of 30 km/sec, whereas comet tails are thrown to the side away from the Sun by the "solar wind," as it is called by the astronomers? Here we have a problem of great, even paramount, importance, which the establishment takes special care to avoid. Some theoreticians have maintained that it is because the Earth has a much greater mass than the comets and therefore a much greater force of attraction. What nonsense! It stands to reason that if the Earth moved through a vacuum (which incidentally is not a vacuum since it is full of particles originating from cosmic radiation), it could not retain its atmosphere while rushing along at such a speed. The fact that its atmosphere is not hurled away in the opposite direction from the Sun, like cometary dust, is an indisputable proof that the Earth and the other planets are drawn along by the ambient medium which compresses them and acts as a protective envelope for them, whereas the comets are devoid of it. The existence of this medium is, moreover, attested by Michelson's negative experiment which could not detect any difference in speed between the Earth and the ambient ether.

It is also evidenced by gravity anomalies consisting of an appreciable diminution in gravity on the north slopes of mountains as compared with the south slopes. These well illustrate that it is the pressure of the ambient medium, being greater on the sun side, which determines both the fall of bodies and the movement of the Earth.

This is further corroborated by the recently made discoveries of equatorial winds rotating faster than the planet. Thus, at an altitude of 150 km above the Earth, there are continuous winds moving at a speed of 40 percent faster than that of the planet's rotation. It is, of course, the ambient medium which causes the heavenly bodies to rotate.

On Venus, which is subject to the tremendous pressure of 95 atmospheres, since it is close to the Sun, the stratosphere rotates 60 times faster than the planet in the same retrograde direction (in 4 to 5 days whereas Venus rotates on its axis in 243 days).

On Jupiter, at the level of the Jovian magnetosphere (95 Jovian radii out), probes have measured a speed of co-rotation of the order of 1000 km/sec. On Saturn, winds reach up to 1770 km/hr at the equator.

The extraordinary speed of Phobos, which revolves three times faster around Mars than the planet does on its own axis, also attests to the influence of the ambient medium.

The same applies to the differential rotation of the Sun, which at the equator rotates upon its axis in 25.2 days, while it takes 26.4 days to do so at latitude 30° and 30 days at latitude 60°. This differential rotation is also appreciable in the case of Jupiter and the giant planets.

The moment one has understood that the heavenly bodies are moved and compressed by the ambient medium, one immediately grasps that it is not hadrons which are responsible for the cohesion of the atomic nucleus, but that it is the concentric pressure to which a star is subject which produces the concentration of particles constituting the atom. These particles having an internal movement (spin) and taking part in the general expansion, probably proportionally to their diameter, the outer particles (electrons) grow larger in the course of time, while the inner, smaller ones, which are in contact with them, revolve all the faster and become smaller the more the outer ones increase in size. Hence the vast difference in volume between protons and electrons.

Attraction having been eliminated, an overall system of dynamics can then be built upon pressure.

THE EVOLUTION OF THE PLANET EARTH

First put forward by Kant and Laplace and still accepted today by the majority of astronomers, the hypothesis of the formation of the solar system by the condensation of a nebula suffers from the major defect of regarding the solar system as an independent unit, whereas it is incorporated in a galaxy whose evolution determines that of the stars forming part of it. Since our galaxy is expanding, the formation of the solar system cannot be explained by condensation. It is at the level of galaxies that the successive phenomena of disintegration followed by condensation take place. The evolution of the solar system is therefore determined by expansion, characterized by the progressive recession of the Sun with respect to the galactic center and of the planets with respect to the Sun which periodically expels them, as well as by ejection of satellites by the planets, also periodic, as soon as they achieve a sufficiently rapid rotation.

Since the condensation hypothesis is incorrect, it obviously cannot account, as we have seen, for the peculiar characteristics of the solar system, such as the existence of planets and satellites, Bode's law, and the differing motions of the satellites according to their distance from the planets. It is equally unable to overcome the difficulties which arise as soon as one wishes to determine the process ending in the formation of the heavy elements which exist on Earth.

The Earth being a planet, geology enables us to discover in close-up, on the ground, the various stages of planetary evolution. It was in 1875 that the Austrian Eduard Suess published a work, *Die Entstehung der Alpen* (*The Origin of the Alps*), in which he stated that the Alps were due to *horizontal* pressures and not vertical ones as in volcanoes, then followed this with a masterly work *Das Antlitz der Erde* (*The Face of the Earth*) translated into French under the title *La face de la Terre*.

The Frenchman Marcel Bertrand recognized the connection which existed between the formation of successive mountain ranges in former times, the most ancient being completely planed down (Huronian, Caledonian, Hercynian, then Tertiary), and wrote (*Oeuvres géologiques* I, p. 18, Paris 1927): "Despite the great irregularities of their contours, the four ranges show a general arrangement of deformations around the pole. They define four continuous or noncontinuous circumpolar zones and constitute the four chapters of the history of the globe . . . But as we know it (the distribution of the ranges) it is enough, in spite of the absence of any geometrical arrangement, to call to mind a connection with the rotation and flattening of the globe."

In 1915, the German Wegener published *Die Entstehung der Kontinente und Ozeane* (*The Origin of the Continents and Oceans*), which was translated into French under the title *La genése des continents et des ocèans* (1925), in which he demonstrated that the American continent, formerly joined to Europe and Africa, became

separated from them during the Eocene (start of the Tertiary) and
drifted westwards, thus creating the Atlantic Ocean which progressively widened.

Finally, in 1922 the Swiss Argand published his work on the tectonics of Asia whose movements he linked with those of the Alpine
range by explaining that the latter had been formed by the collision
of the European continent and the African basal mass (socle), whilst
the Himalayan folds resulted from the overriding of Gondwanaland
(especially India) by the Asiatic socle.

Such movements discovered by these pioneers, who were joined
by an international élite called the "mobilistes," must be explained.

Too often there is a tendency to forget that the Earth is rotating,
and this rotation has its effects, as Marcel Bertrand so accurately
observed. At present it rotates on its own axis in 24 hours, which
leads us to believe that it was always thus, whereas, in fact, this rotation has evolved since it is determined by the movement of the ambient medium. Furthermore, the Earth is the first planet out from
the Sun which has a satellite, which is not an occurrence without
consequence. After it had been expelled from the Sun, it passed
through the stage of Mercury, then through that of Venus, whose
slow rotation is now established. Geology teaches us that rock formations prior to the Primary era do not contain different layers,
except in the intermediate regions between day and night as a consequence of the inclination of the axis of rotation. In the earliest
times, therefore, it always presented the same face to the Sun. When
it began to slowly turn on its own axis, one hemisphere was lit by
the Sun for thousands of years. The succession of fundamentally
different layers from the time of the Primary era reveals this slow
rotation and explains the succession of climates essentially determined
by longitude, and not by latitude, as is the case today. For a given
region, the alternation of tropical climates and glacial periods produced, in addition, a periodic reversal of the magnetic field, which
is evidenced in the rock formations.

As it moved away from the Sun, the Earth reached a zone which
was less compressed by the ambient medium (probes have established
an atmospheric pressure of 95 atmospheres on Venus). It then began
to rotate a little faster on its axis and to flatten out. The equatorial

bulge resulting from this readjustment ended by being expelled and forming the Moon. The date and location of the expelled part may be reconstructed through the upheavals which preceded and followed the expulsion. The date is certainly a little prior to the breaking up of the continents described by Wegener and therefore between the end of the Secondary era and the beginning of the Tertiary. As for the location, the Moon was certainly in the zone now occupied by the Pacific trough, the deepest on the globe and the most unstable, as attested by the ring of fire which surrounds it, still characterized today by volcanic eruptions and earthquakes. The readjustment has therefore not yet finished.

The expulsion led to a change in the dynamic equilibrium and in the axis of rotation. The waters which invaded the trough have a density three times less than that of rock. It is in the wake of this occurrence that the continental drift described by Wegener took place. While America separated from Europe and Africa and drifted westwards towards the Pacific trough, Europe, Africa and Asia were traveling in the opposite direction, also towards the trough. The former seam between the continents is still visible today along the great median submarine range, parallel to the coasts, which divides the Atlantic from North to South.

Whereas until the Tertiary the main mountain ranges seem to be the result of the flattening of the terrestrial globe, continental drift, via abrasion with the subjacent rock formations, is causing the creation of vast ranges in front of these great rafts, the most characteristic example of which is that of the Rocky Mountains and Andes.

The Moon's expulsion thus had the effect of breaking up the Earth's crust into plates whose movement has not yet ended. Their collision zones, which are the cause of earthquakes and volcanic eruptions, are studied by specialists in plate tectonics. Unfortunately, these specialists who attribute these displacements to convection currents, which is not an explanation, seem to be unaware of the effects of rotation and the general expansional evolution of the planet.

There should be no misunderstanding: evolution is not limited to the past. It is a continuous phenomenon which applies to the present and will continue in the future. It is increasingly deforming the rotating Earth whose equatorial diameter, which is 43 km greater than between the poles, is the sign of an advanced process. The deformation, which can only become more marked, will inevitably result in the expulsion of a second Moon. Until this upsetting phenomenon occurs, grave events will rouse humanity and perhaps—who knows?—awaken dormant minds. It is about time that it was appreciated that the Earth is a wandering body whose evolution is occurring at great speed and that it would be preferable to seriously study this fundamental phenomenon rather than search the sky for mythical black holes.

We have explained and illustrated this evolution in *La Terre s'en va* (*Earth's Flight Beyond*) and consider it pointless to repeat ourselves. Our aim is to show that by ridding ourselves of all preconceptions, as Descartes enjoins us to do, and by taking his view of the universe as a basis, science and common sense may be reconciled and a true picture obtained of the future development of our planet, a picture which prompts us to reflect and prepare ourselves for coping with the increasingly serious trials to come. So long as the broad outlines of evolution are not taken into consideration, we shall never be ready in the event of catastrophes and will be content to bury the victims and send medicines and blankets for the injured and homeless. These expedients seem to us distinctly inadequate.

7

Open Minds

After more than a quarter of a century of explaining my thesis of the continuum of the universe, integral dynamism and universal evolution, I invite thinkers to reflect upon the causes and effects of this evolution which extends from the infinitely small to the infinitely large and to realize that the privilege of permanence which we believe we deserve, is no more than illusion.

Some scientific personalities have been well disposed to pay attention to these ideas. The late-lamented Professor Pasteur Vallery-Radot, member of the Academy of Medicine and the French Academy, wrote this to me in connection with *L'Univers en marche*: "You have written an astonishing book giving an insight into the whole universe. It is a magnificent, personal synthesis. It will arouse the curiosity of the young and cause those in their declining years to reflect. I do not think that such a wide-ranging work has been attempted since the 18th century. I warmly congratulate you."

On the subject of *La Terre s'en va*, in which I studied the consequences of this general expansion for our planet, he expressed himself thus: "What an enthralling book *La Terre s'en va* is! I have read it twice and written notes on it. I know of no pages more fascinating than *those where you allude to the great events about to take place. The satellite which you envisage gives cause to ponder. Here is a book which restores Man to his proper place in the Cosmos*."

His interest in the books I have written was constant: "Few recent books have aroused my interest as much as *Méditations sur le mouvement.* This is a *great* book which due to the author's remarkable insight is very thought-provoking."

"*L'évolution universelle* is disturbing. You wonder what the universe is coming to, where the Earth is going, where life and thought are leading. Here is a book which makes one think."

Thus the heir to that open-mindedness which enabled Louis Pasteur to make his wonderful discoveries did not need a drawing to understand that within an expanding galaxy, the solar system itself was also expanding. And the venerable professor took the trouble to read *La Terre s'en va* twice, he found the subject so fascinating.

Other open-minded people have taken the trouble to reflect upon the theses I constantly put forward. Thus a member of the Academy of Science wrote the following in connection with my work entitled *L'imposture scientifique*: "Because of its title, quite shocking when applied generally, I should doubtless never have bought your book. Having skipped to the end of it as always, I have to thank you most sincerely for having made me a present of it, thus enabling me to become acquainted with your philosophical ideas. This is a time in my life when I am revising my own ideas on the great problems, great sceptic that I have always been. Thank you once again."

Another member of the Institute wrote to me "that he had read this very original work with great interest. He was also disturbed at the questioning of so many fundamental ideas which he thought were almost conclusive. Should we live thus in a kind of chronic, clandestine sophism?"

Another opinion of a member of the Academy of Science: "Thank you very much for having sent me your book *L'imposture scientifique* which I hold in high regard. As a biologist, I would be even more severe than you with the errors which you expose. They conceal our ignorance with a conformity which it is strange that men of science are unwilling to abandon."

But while open minds are taking the trouble to reflect upon the expansion of the galaxy and its effects upon the solar system, the anomalistic year, the law of Bode, the presence and motion of the satellites, the atmosphere which remains around our globe despite its fantastic speed of travel around the Sun, these open minds are not legion and form a brilliant but small minority by comparison with those minds which are closed.

8

Closed Minds

If many scholars know almost everything about a particular subject but understand almost nothing about the world in which they live, it is because they are steeped in prejudices and avoid the basic problems as soon as they begin to bother them.

In connection with *La Terre s'en va*, on October 20, 1958, the *Revue des questions scientifiques* published the following critique:

> The author, who is convinced that the men of the 6th century before Christ were more acquainted with the catastrophes which struck . . . humanity . . . throughout the Quaternary era . . . than we are today, refuses to allow mathematical formulae as explanations; he rejects Einstein and even Newton and takes as the point of departure of his cosmology the idea of a filled space, constructed upon Descartes' model of vortices. This does not prevent him from having recourse to radioactivity, the expansion of the universe and sputniks, which presuppose Newton and Einstein, an empty space and the mathematics of calculus, but basically what does it matter? Since this book is a "protest against the tyranny of official science," why should the author submit himself to the only tyranny of science which is that of logical coherence and yielding to fact? Scholars who need no lessons in non-conformity from anyone, but know how difficult it is for them to get their results published, will, however, ask how such works find a publisher and purchasers?
>
> P. de Béthune

The end of the article deserves our full attention, since it clearly explains the pretension of some official circles to the exclusive rights over the truth and their relentlessness in preventing the free discussion of ideas by disparaging not only authors but publishers who have the insolence to publish texts written by non-conformists. What a terrible crime—to disseminate a work explaining that the solar system cannot escape universal evolution! What stupidity—to explain Bode's law by evolution rather than by chance! And what impertinence to attribute the shifting of the aphelion to the enlargement of the Earth's orbit rather than to construct an anomalistic science! In short, in the mind of this critic with the aristocratic name who must obviously consider Descartes to be a moron and the photographs of spiral nebulae to be the product of faulty equipment, logical coherence requires that the freedom of the press remain a monopoly of the objectors to the idea of evolution.

Other dull minds, impervious to the idea that the solar system cannot escape universal evolution, attack the booksellers who disseminate new ideas. Thus, a professor of the Paris Science Faculty, the author of a book entitled *Origine et évolution des mondes* (*Origin and Evolution of Worlds*), wrote this to a bookseller who had sent him a prospectus on *La Terre s'en va*: "It is a sheer scandal that a firm as reputable as yours should distribute a book as inept as M. Jacot's. This work has nothing whatsoever to do with science and is the most extraordinary farrago of erroneous ideas that I have encountered to date. *I hope that you will take the action appropriate to this protest* and remain yours . . ."

Thus this reverse evolutionist who has not understood galactic expansion at all and for whom "it is not idle to assert that with suitable improvements the nebula hypothesis is the only one which may account for the formation of the solar system" (*op. cit.*, p. 330) is not content with expressing his disapproval—which is his right—but demands that the publisher "take the action appropriate to his protest." And this on a letter bearing the heading: Ministry of National Education—National Center of Scientific Research—Institute of Astrophysics. What openness of mind for a professor, and what courage to address himself to the publisher rather than to the author!

Some professors do not hesitate to condemn you without properly reading your thesis and make no attempt to conceal it from you. Thus, a professor of physics at the University of Neuchâtel, to whom I had sent a complimentary copy of *L'imposture scientifique*, answered me on April 18, 1974, as follows:

> "It would be wrong to say that I have thoroughly read through your book *L'imposture scientifique*, which I am grateful to you for sending me . . . When you claim to replace physical theories inspired by experiment and capable of explaining, partially and temporarily perhaps, but effectively and in detail, with a crude and singularly naive conception of 'atomic gearwheels,' without doubt you exceed the limits of temerity . . . There is no doubt that you have an easy and striking style. This should not disguise a certain poverty of 'scientific inspiration.' "

Thus, without taking the trouble to follow the logical development of my thesis regarding the behavior of forces, a professor of physics condemns the poverty of my "scientific inspiration" out of hand. Is he unaware that Max Planck and Niels Bohr hold very different conceptions of atomic structure? Has he never considered the significance of spin within particles, and does this rotational movement within particles mean nothing to him? The eminent professor, wishing to show his openmindedness in his own way, ends his letter thus: "If despite the negative comments that even the *partial* reading of your book compels me to make to you, you think that a discussion would be useful, I should most willingly spare a few hours on it with you."

What condescension! I was so touched by it that I replied as follows: "You are too kind. Considering the vastness of your knowledge and the poverty of my inspiration, I think that a discussion between a supporter of static and invariable physics and the supporter of an evolutionary physics could only end in a dialogue of the deaf. It would be ungracious of me to inflict such a disagreeable experience upon you."

The few extracts reproduced above are only a small sample of the hostility encountered by the ingenuous person who thinks he should put forward non-conformist ideas. Indifference is an obstacle even more difficult to break through than hostility. On February 8, 1979, I sent the Paris Academy of Science a paper entitled: *Le système solaire est-il en expansion?* (*Is the Solar System Expanding?*) in which I set out the principal facts supporting the thesis of expansion. The honorable, permanent secretaries did me the signal honor of acknowledging the receipt of this paper and of informing me that they had deposited it in their files where it rests in peace. In principle, I have nothing against the sleep of the just, but I do not think that this is the best way of solving the great problems of the universe.

At the end of 1979, I sent the manuscript of the present work, then entitled *Retour à Descartes et à la logique en science* (*A Return to Descartes and to Logic in Science*), to ten of the best-known French publishing houses. It was rejected by them all. Most of them did so very politely, almost apologizing for not having a series devoted to the expansion of the solar system. Some seemed to be watching the pennies. Thus Buchet-Chastel replied that my manuscript would be returned after I had sent the costs of postage. In order not to pay them, Fayard simply returned the manuscript. This kind of courtesy did not unduly surprise me, as I have had some experience with these purveyors of others' thoughts.

In order to highlight the importance of quanta, discovered by Planck, for the formulation of an overall theory regarding the evolution of the universe, I summarized the preceding work and on October 10, 1980, sent a paper to the Academy of Science entitled *Realité et dynamique de l'éther* (*Reality and Dynamics of the Ether*). On November 19, 1980, the permanent secretaries returned my manuscript attaching an expert's report which stated that "the comments it contains are purely qualitative. But science can only be quantitative." I had always believed that the academy was of the Platonic school, made famous by its writings on the world of ideas, and I thought that the members of the academy, faithful to this tradition, were given to confronting ideas, for how could figures mean anything without an idea? Yet here was the Academy of Science of Paris writing to me that my manuscript did not deal with ideas or intellect. To be sure, my concepts with regard to the evolution of the solar system do not attain the level of the mathematical expansions on black holes. I admit also that an academician's armchair is for sitting on and not made for moving, but it seems to me that the problem of the future development of our planet should merit more than the indifference of the leading lights of French science. In order to be "Cartesian" in complete independence, Descartes left France to go to Holland and then Sweden. On his death there was a dispute over his remains. His headless body was brought to France sixteen years later. At the funeral ceremony in 1667, the funeral oration was forbidden. Finally, his head was brought back to France in 1822, but must the spirit of Descartes be banished from it forever?

However, the Cartesian spirit was honored once more a little while ago at the Academy of Science. Louis de Broglie, Nobel Prize winner and former permanent secretary of this academy, a specialist in wave mechanics, wrote this in the preface to his work *Nouvelles perspectives en microphysique* (*New Perspectives in Microphysics*), published in 1956:

> After the birth of wave mechanics, I tried for several years, from 1923 to 1927, to find an interpretation consistent with the idea of causality, using, in the tradition of physicists, a representation of physical reality with the help of precise images within the framework of space and time. The difficulties that I encountered in making this attempt, the hostility that it aroused on the part of other physics theoreticians, led me in 1928 to abandon it and for close on 25 years I adopted the probabilistic interpretation arising from the works of Messrs. Born, Bohr and Heisenberg which has become the official doctrine of theoretical physics . . . Those who are interested in the psychology of scholars will be happy to learn that I hereby give a few hints on the circumstances which have provoked in my mind an unexpected return to those long abandoned ideas . . . I shall not speak of the difficulties that I have always experienced in putting forward the probabilistic interpretation of wave mechanics, of the regret that I have often felt in seeing the authors forget or belittle the physical intuitions which were the origin and basis of this profound theory, nor *of a certain secret hankering after Cartesian clarity which came over me when I tried to find my bearings in the midst of the fog in which present-day quantum physics is shrouded.*

Later, in the chapter entitled "Contemporary atomic and quantum physics," which appears on pages 129ff of the *Histoire générale des sciences* (*General History of the Sciences*), Vol. 3, II, "The 20th Century," de Broglie states: "Quantum field theory is prohibited, 'a priori,' from considering a pin-pointing of the corpuscles (sic) and attributing a dimensional structure to them, as a consequence of defining concepts similar to those of the "electron radius" in Lorentz's classical theory. Personally we think that the idea of the localization and dimensional structure of corpuscles should in some way be re-established."

Louis de Broglie continues: "In view of all these difficulties of interpretation, it seems to us that contrary to what some misinformed persons may think, theoretical physics is at present going through a period of great stagnation. Even the warmest supporters of its present direction acknowledge this, and one of them, R. Oppenheimer, recently wrote: "It is obvious that we are on the verge of a very important, probably very heroic and in any event totally unpredictable revolution in our interpretations and our theories in physics."

Oppenheimer, who was the creator of the atomic bomb and who was certainly acquainted with Einstein's theories (it was Einstein who recommended him to the president of the United States for manufacturing the bomb), realized the inadequacy of these theories and knew perfectly well what he was talking about when he announced the necessity of a very serious and heroic revolution in the interpretation of physical phenomena. Are the heroes of modern physics too tired to carry out this revolution, and do they think that it can only take place through the quantitative? There can be no revolution of theories without a complete revision of ideas, without a return to Descartes, to his vortex theory and to his logic, which adheres to his reality. This return compels us to abandon the preconceptions of attraction and stability to replace them with the sound idea of a continuous universe, governed by pressures resulting from particles whose essence is movement, which causes instability and then an evolution extending from the infinitely small to the infinitely large.

9

If England Were an Island,
We Should All Know It!

"If England were an island, we should all know it!" said some
anonymous person who took his modest intelligence for intuitive
knowledge. "If the Earth turned around the Sun, we should all know
it!" they said at the time of Copernicus. "If the Earth were moving
away from the Sun, we should all know it!" now consider those
minds dominated by closed ellipses.

In order to find out whether a piece of land is an island, one
must have an overall view of it or make a circuit of it. Thus it is
with all problems. For a long time it was believed that the Earth
was the center of the universe; then, thanks to Aristarchus of Samos,
Copernicus and Galileo, Man established that the Earth is a planet
which revolves around the Sun. It is a spacecraft which is rushing
along at the fantastic speed of 30 km/sec.

But who comprehended the full implication of Copernicus'
discovery? Who understood what a planet really is? Where it comes
from? Where it goes to? What it becomes? Whence it derives its
movement? What are the consequences of the movements which
animate it? While the astronomers effectively altered the coordinates
of the solar system after Copernicus' discovery, nevertheless the majority have remained prisoners of medieval geocentrism, clinging
doggedly to permanence and remaining impervious to the idea that
the solar system cannot escape universal evolution.

To be sure, it is more convenient to put problems on the basis of invariance and to consider time as an invention of Man who can manipulate it and rewind it as he pleases. But, far from being a peculiarity of perpetual repetition, an exception to the general principle of invariance or an accident due to chance which affects certain atoms, *evolution is the fundamental phenomenon which, by acting within each particle, governs the whole universe and must serve as the basis of all sciences.*

A physicist can never formulate an exact science of physics if he regards his standards of space and time as being stable when they are, in fact, evolving, if he is unaware that the universe is continuous and dynamic in each of its parts and if he fails to recognize that the universal substance, which constitutes both the matter of stars and that of interstellar space, is constantly evolving because its essence is movement.

Since evolution is a universal phenomenon which is continuous in space and time, all cosmologies, ancient or modern, founded upon the preconception of an initial phenomenon called the creation of the universe, followed by an indefinite permanence, are as untenable as the pretension to exclude the solar system from the expansion of our galaxy.

Similarly, in order that a geologist may correctly interpret his findings in the field, he must take account of the evolution of both the Earth's rotation and its translational movement, in particular of its slow rotation in former times. By plunging one hemisphere into darkness for thousands of years and subjecting the other to the incessant heat of the Sun's rays, it produced climatic conditions which were completely different from those at present, a succession of emergences and submergences of the land, as well as an alternation of the Earth's magnetic field.

The biologist who wishes to understand the development of life on our Earth cannot disregard the circumstances under which life evolved in the past nor be unaware that the evolution of species is an evolution of the second degree superimposed on that of matter. One cannot, in fact, superimpose evolution onto permanence. And how could people concerned about the future of our planet form an accurate idea of the future if they have not grasped the mechanism of evolution, its causes and effects?

Is it too much to ask to suggest this change·in outlook, to clear one's mind and drive out the prejudice of permanence? The effort is considerable, I agree; but if, by passing from invariance to evolution, from the closed circuit to the open spiral, we pass at the same time from closing to opening our minds, then why not try it? Haven't we all the more grounds for doing so since they are not theoretical discussions which are involved but problems of incredible implications?[1]

Before the all-powerful Inquisition, Galileo had to admit on his knees that the Earth was the immovable center of the universe. Legend has it that after he rose to his feet again, he exclaimed: "Yet it does move" (*Eppur si muove*).

Today, before the omnipotence of the prejudice of permanence which has determined that the Earth remains at the same distance from the Sun each year, one can only exclaim: "Yet it does move away." Even if it entails seeming like an idiot in the eyes of short-sighted scholars.

---

1. An examination of the implications of evolution for the destiny of our planet would justify a lengthy exposition. But we cannot repeat here what we have already set out in *Earth's Flight Beyond* and in our other works.

APPENDIX

APPENDIX

Comparison of the Assertions of Official Science with Reality

If we summarize this great clash of ideas regarding the structure of the universe and its evolution, we should find that a critical mind can accept neither the ambiguities nor the contradictions, which make the universe incomprehensible, nor the deficiencies of a science unable to explain the most outstanding phenomena of the solar system, such as Bode's law on the distances of planets from the Sun, the shift of the aphelion in the direction of travel, the apparent shrinking of the Sun's diameter, the Moon's recession as well as other features (movements and volume of the planets, existence, number and position of the satellites, the mystery of Phobos, etc.).

As long ago as 1944, in *L'univers en marche*, then in various other works, I disputed the validity of the most mistaken assertions of official science, all tainted with the prejudice of permanence, and demonstrated that they were manifestly contrary to reality as characterized by the continuum, integral dynamism and universal evolution. The principal ones from among these mistaken assertions are as follows:

MISTAKEN OFFICIAL ASSERTIONS	REALITY
1. The universe is discontinuous, for the most part empty, so that the ether does not exist.	1. The universe is continuous thanks to the ether which fills up all of interstellar and intergalactic space.
2. If the ether exists, it must be deprived of any mechanical	2. The ether manifests itself through its capacity to act in

103

MISTAKEN OFFICIAL ASSERTIONS   REALITY

property, in particular that of movement (Einstein).

wavelike fashion and thus through its movement. To deny this is to deny the existence of hertzian waves (radio, television, etc.). This is to deny the evidence. In order to act in wavelike fashion and transmit movement, the ether must be composed of dynamic particles actuated by an internal movement. Furthermore, these moving particles form vast vortices, as Descartes described and as attested by the photographs of spiral nebulae.

3. The heavenly bodies move through space under their own energy and are attracted to one another in direct proportion to their mass and in inverse ratio to the square of their distances (Newton).

3. The heavenly bodies are drawn along like flotsam by the ambient medium, the ether, which compresses them and imparts both their rotation and translation.

4. The solar system has its origin in the condensation of a nebula.

4. Such a condensation is incompatible with galactic expansion. Originating from the center of our galaxy, the solar system is progressively moving away from it at the same time expanding.

5. The creation, now completed, is followed by a perpetual recommencement governed by the attraction which causes the planets to describe closed ellipses which keep them at the same distance from the Sun through the ages.

5. There is no completed and inexplicable creation, but a continuous, perpetual and universal evolution, in which the solar system takes part; the latter's expansion causes the planets to describe spirals which progressively remove them away from the Sun.

Mistaken Official Assertions	Reality
6. The geometrical progression with a common ratio of 2 which repeats itself seven times and distinguishes Bode's law for the distances of the planets from the Sun is explained solely by chance.	6. Bode's law is the broad law which attests to the evolution of the solar system. Its geometrical progression with a common ratio of 2 indicates that, during the same time, the centrifugal force proportional to the radius which exists in the central region of the solar system moves away the planet located at distance 1 by a distance of 1 causing it to pass to distance 2, removes by 2 that at 2 which passes to 4; that at 4 moves away by 4 and passes to 8, etc.
7. The fixed number to be added to Bode's law has no meaning.	7. This fixed number means that at the time of its expulsion by the Sun the planet cleared a very great distance in one bound.
8. Inexplicable also is the fact that the geometrical progression ends beyond Uranus, where the planets are approximately 1500 million kilometers apart from one another.	8. The geometrical progression ends at the point where the centrifugal force ceases to be proportional to the radius when it causes the reversal of the vortical movement which passes from positive to retrograde motion.
9. Again inexplicable is the position of Uranus which at each solstice presents one of its poles to the Sun, its axis of rotation being almost parallel to that of its orbit instead of being almost perpendicular as in the case of the other planets.	9. Uranus lies precisely in the zone where the change in direction of the solar vortex causes the planet to tilt over so that its South pole becomes its North pole and vice versa.

MISTAKEN OFFICIAL ASSERTIONS   REALITY

10. The extraordinary speed of Phobos, which revolves three times faster around Mars than the planet rotates about its own axis, is incomprehensible.

10. This exceptional speed, incompatible with Newtonian mechanics, results quite simply from the recent expulsion of Phobos by Mars.

11. The innumerable striae shown in the photographs of Phobos taken by NASA are inexplicable.

11. These grooves result from Phobos' abrasion against Mars at the time of expulsion.

12. The particularly eccentric orbit of Mercury is unexplained.

12. This exceptional eccentricity is the effect of the recent expulsion of Mercury by the Sun.

13. Newtonian mechanics does not explain why the satellites of Jupiter and Saturn which are close to the planets revolve in the positive direction, and those farthest away in the retrograde direction, while in the case of Neptune the reverse is true.

13. These differences in motion are the natural consequence of the change in direction of the whole vortical movement starting from a critical distance from the center, as a result of centrifugal force.

14. Finding that the Sun's diameter seems to be contracting by 0.1 percent per century, and that if this tendency should continue at the same rate the Sun's diameter would be practically zero at the end of 200,000 years, the astronomers, conscious of this absurdity, assume a reversal of the tendency at the end of a certain time.

14. Contrary to overall expansion, there is no contraction of the Sun's diameter. This is nothing more than an illusion due to the progressive recession of the Earth away from the Sun.

15. The year marked by the return to the aphelion (the far-

15. This lengthening of 4 minutes and 45 seconds is due to

MISTAKEN OFFICIAL ASSERTIONS	REALITY
thest point of the Earth's orbit from the Sun) which is 4 minutes and 45 seconds longer than the sidereal year (complete revolution) has been called "anomalistic" because this lengthening was incomprehensible.	the spiral described by the Earth which is diverted forwards and outwards as a result of the centrifugal force existing within the solar vortex.
16. The continuous recession of the Moon away from the Earth is attributed to the tides.	16. This phenomenon results from the overall expansion which is progressively moving the planets away from the Sun and the satellites away from the planets. It proves the expulsion of the Moon by the Earth.
17. Hubble's law regarding the "flight" of the galaxies is interpreted as meaning a growth of the void between the galaxies.	17. Hubble's law, which states that the galaxies are moving away from each other at a speed proportional to the distance separating them from the Earth, only has meaning within the concept of a continuous universe, since the Earth is not its center. It is incompatible with Newton's law of attraction.
18. The theory of the "Big Bang" at the beginning of the universe presupposes the existence of another universe into which expansion occurs. It is therefore unacceptable. It does not explain what happened before the "Big Bang." Assuming a beginning, it implies an ending.	18. The theory of universal evolution according to which new galaxies are formed at the expense of those which break up is free from these contradictions. It excludes a beginning and an end of the universe, each as inexplicable as the other.

MISTAKEN OFFICIAL ASSERTIONS   REALITY

19. The Earth has always revolved on its own axis in 24
hours.

19. The Earth's rotation has
evolved through the ages and
continues to evolve, just as its
translational movement does.

20. The alternation on Earth of
tropical and glacial periods is explained by variations in solar
radiation.

20. The alternation of tropical
and glacial periods as well as the
longitudinal distribution of these
climates (and not latitudinal, as
at present) are explained by the
slow rotation of the Earth which
exposed one hemisphere for
thousands of years to the Sun's
rays, whilst the other was plunged
into night.

21. The alternation of the
Earth's magnetic field through
the ages, verified in rock formations, is due to a periodic change
in the position of the poles, the
North pole becoming the South
pole and vice versa.

21. There is nothing to support
a *periodic* change in the position
of the poles. The alternation of
the magnetic field is explained by
the slow rotation and continuous
exposure of one hemisphere to
the Sun's rays.

22. The disappearance of various species in certain formations
and their reappearance in higher
formations thousands of years
later are unexplained.

22. They are explained by slow
rotation which allows some
species to travel in the opposite
direction to the rotation in order
to remain in a favorable climate
and to reappear above formations formed in an unfavorable
climatic zone.

23. Even though comet tails are
repulsed from the Sun by the
"solar wind," the Earth's atmosphere remains around the
globe, although the Earth travels

23. It is pretty obvious that the
attraction assumed by Newton
would be quite incapable of
keeping a gas around a heavenly
body which speeds along at

MISTAKEN OFFICIAL ASSERTIONS

REALITY

at 30 km/sec along its orbit. According to the official astronomers, it is thanks to attraction that the Earth keeps its atmosphere.

30 km/sec. If the atmosphere remains around the Earth, it is because it is compressed by the ambient medium, the ether, which draws our planet along in its course around the Sun and causes it to rotate as is shown by the equatorial winds rotating faster than the planet.

24. Newton postulated the existence of attractive forces and maintained that instantaneous attraction took place outside of time. According to some of his followers, attraction would act in the form of gravitational waves.

24. No attractive force has been discovered and its mechanism has never been demonstrated. The G.C.W.M. defined force as resulting from a pressure. A wave being the product of an impact upon a mass capable of undulation, waves cannot attract not having the capacity to recede.

25. According to Einstein, Michelson's experiment demonstrates the constancy of the speed of light. He asserts, in addition, "that it is more natural to represent physical reality as a four-dimensional being instead of representing it, as has been done until now, as the evolvement of a three-dimensional being."

25. Properly understood, Michelson's experiment attests the accuracy of the wave theory of light, according to which interferences should not occur, and confirms that the ether is not immobile, but draws the Earth along its orbit. To deny evolution is to deny the past and the future; it is to talk nonsense. Einstein would have been well advised to have meditated upon the propositions of Heraclitus: "Nothing is, everything is becoming," and of Plato: "knowledge is a probable opinion of becoming."

MISTAKEN OFFICIAL ASSERTIONS	REALITY
26. Planck's constant implies the discontinuity of the universe.	26. Planck's constant means that the universe is formed of minute, dynamic particles pressed against one another by their internal movement which impels them to expand, with the result that the universe is continuous.
27. According to circumstances, light is sometimes a flow of corpuscles traveling through space, like rifle bullets, sometimes a wave motion of the ether on the spot.	27. The phenomenon of light cannot have two contradictory mechanisms. It is always produced by a material impact upon the ether which transmits this impact in the form of wave motion. Only the wave theory of light is valid.
28. The speed of light is that of the actual displacement of particles.	28. The speed of light results from the *internal* movement of the particles which transmit the impact. Thus it is of the same order of magnitude as that of electricity in good conductors.
29. Atoms are formed of particles grouped in a nucleus around which the electrons revolve in various orbits like the planets around the Sun (Bohr) or like a perfectly symmetrical ring rotating about itself (Planck).	29. Formed of particles having an internal rotational movement (spin), the components of an atom can only behave towards each other like cogwheels. When in contact along their equators, they form couples revolving in opposite directions from one another. When in contact at their poles, they form axes turning in the same direction. Atoms are assemblages of axes and couples. The central particles, which are more compressed than those on the periphery, are smaller and revolve more rapidly.

MISTAKEN OFFICIAL ASSERTIONS	REALITY
30. The mass of electrons is much less than that of protons.	30. The mass of electrons is necessarily the same as that of protons since these particles when in contact are continually exchanging their movement and, therefore, their mass. The difference which appears to result from various experiments is merely apparent and arises from the greater volume of electrons and their lower density.
31. There are two categories of atoms: stable ones, which do not evolve through time, and radioactive ones, which do evolve.	31. All atoms evolve with expansion. In simple atoms, this results solely in an increase in diameter. In complex atoms, expansion compels some particles to alter position within the atom. Since those which are in contact along their equators revolve in the reverse direction, while they turn in the same direction when they are superposed, as soon as they reach 45° to the plane of their orbit the incompatibility of movements causes expulsion.
32. According to the chemists, atoms form molecules by interlocking with each other, one peripheral electron taking the vacant place of another atom which has less than eight peripheral electrons. But this concept contradicts that of the physicists according to which electrons orbit around the nucleus like planets around the Sun. Such a movement would preclude any interlocking.	32. The chemists' concept, whose validity is proven by all experiments on the arrangements of atoms in molecules, confirms the thesis by which atoms are assemblages of cogwheels.

MISTAKEN OFFICIAL ASSERTIONS	REALITY
33. A force field is continuous and therefore cannot be formed of distinct particles.	33. A force field can only be an assemblage of actual units of force which are in contact and continually exchanging their motion.
34. There is a fundamental antithesis between matter and ether according to the physicists who accept the existence of the ether.	34. There is no antithesis between ether and matter, both being of the same substance whose essence is movement. There is no substance without movement nor movement without substance. In fact, matter is nothing more than a condensate of ether, which is the universal substance. Ether and matter are interdependent through their permanent contact. Since all movement involves change, both take part in the universal evolution which affects both matter and interstellar and intergalactic space. Ether may be transformed into matter and vice versa.

We could extend the list of contradictions and ambiguities, but one does not need to be very discerning to realize that the reality is logical and that establishment science is not. For it to become so, it would be enough for it to accept the reality of the ether and its dynamic properties. But how many establishment scholars are prepared to abandon their prejudices?

The Paris Academy of Science does not seem at all prepared to abandon its prejudices since their expert, who was entrusted with scrutinizing my text, wrote thus: "It is, of course, vital that a theory be coherent and it is necessary that it be subjected to criticism of an epistemological nature. But this criticism must not throw out the baby (Newton?) with the bath water (the sea of prejudices!)" Amen.

Postscript

This book was written some months ago when, at the end of August 1981, NASA's *Voyager 2* took some remarkable photographs of Saturn, its satellites and its rings. These photographs are exceptionally valuable in that the successive shots allow the development of the phenomena taking place around the planet to be followed and attest on the one hand to a "gravitation-repulsion," i.e., the opposite of Newtonian attraction, and on the other hand to the spirated development of the rings, i.e., their progressive movement away from the planet.

These findings thus confirm the thesis that we have put forward in our works since as long ago as 1944, in *L'Univers en marche* and in more detail in *La Terre s'en va*, namely that the planets are moving progressively away from the Sun, which periodically expels one of them as a result of its rotation, and that the satellites are themselves also moving away from the planets from which they are periodically expelled in the form of compact units or debris forming rings.

The continuous deformation of a rotating mass in the form of a flattening of the poles and equatorial swelling is an indisputable phenomenon. In his works, Henri Poincaré depicted this successive deformation and its final culmination in the expulsion of the part of the mass farthest from the center of gravity. Furthermore, logic compels the idea that within an expanding galaxy, the constituent parts, of which the solar system is one, cannot be governed by permanence, but take part in this expansion.

The photographs of Saturn taken by NASA do not relate to one special case but represent the overall law applicable to all the planets, so that as new photographs come to be added to the old, men of science must, sooner or later, abandon the Newtonian theory of gravitation by attraction and adopt the vortical, evolutionist view of our solar system.

Under the title "The New Milky Way" (*Science* No. 4603, June 17, 1983), Leo Blitz, Michel Fich and Shrinivas Kulkarni wrote that "our understanding of the large-scale structure of the Milky Way has undergone considerable revision during the past few years. The galaxy is larger and more massive than previously supposed; the newly discovered mass consists of non-luminous matter which is likely to be the dominant form of the universe . . . An understanding of the universe as a whole, as well as the origin and formation of galaxies, demands an understanding of this new and strange component of the Milky Way, and it is sure to be the major challenge for astronomy in the coming years."

Same opinion in "The Rotation of Spiral Galaxies," by Vera C. Rubin (*Science* No. 4604, June 24, 1983): "Much as 90 percent of the mass of the universe is presumed to be responsible for the high velocities of stars and gas in the disk of spiral galaxies . . . Future astronomers will have to be clever in devising detectors which can map and study this ubiquitous matter which does not reveal itself to us by its light."

It allowed me to send to *Science* an article entitled "What Is a Non-luminous Mass and How Is It Acting?" in which I declare that two kinds of light cannot exist, one which undulates on the spot and another which travels through the empty at the speed of 300,000 km/sec. The only accurate theory of the light is that of the undulation of a non-luminous mass—the ether—which produces the luminous phenomenon when its undulation comes in collusion with material bodies. The hertzian waves prove the existence and the movement of this ether, invisible but incontestably dynamic.

The light cannot serve to settle the mass. The empty does not exist in the universe which is full, full of the dynamic mass of the ether, of which the mass is much more important than that of bodies (and gas), so that it is not the celestial bodies which attract the environment but it is the dynamic ether which determines the movement of galaxies, stars, planets and satellites, their translation, their rotation and their evolution. All the substance of the universe is moving and dynamic, but the non-luminous ether is the principal motor of the universe. This new cosmology gives an entirely different image of the structure and the evolution of the universe than the official cosmology. I thought that a serious confrontation of both cosmologies would be beneficial for the knowledge of our destiny. But the editors of *Science* refused to publish my article.

When in the year 1543 Copernicus announced that the Earth turned around the Sun, the Inquisition condemned this thesis as an insult to God and the human dignity. Giordano Bruno, who had the impudence to adopt this idea, was burned with his books in the year 1600. Great ideas cannot enter in narrow minds, and ninety years after the death of Copernicus, Galileo Galilei was constrained to abjure this "heretical and absurd doctrine."

When in the year 1983 an impertinent author writes than in an expanding galaxy the solar system is also expanding so that the Sun expels periodically a planet according to Bode's law and that the planets expel also periodically a satellite or a ring in consequence of their expansion and rotation, the Inquisition of *Science* finds this idea so heretical and indecent that it seems unworthy of being published.

By the hard Inquisition of the sixteenth century the public had the possibility to compare the ideas which were condemned while by the soft Inquisition of the twentieth century the ideas condemned to silence have no possibility to be known and discussed.

The official scientists continue to teach that a Newtonian "attraction" of 9.80 m/sec at the surface of the Earth can maintain the atmosphere around our planet, which is running at the speed of 30 km/sec. And they will continue to teach these old strings of Newton unless . . . perhaps . . .

Notes on the Illustrations

THE UNIVERSE

Cosmology is only valid if it is based on fact. The following illustrations represent, on the one hand, the way in which Descartes saw the world and, on the other, the facts which show that this concept, with the addition of some recent discoveries and slightly amended, does correspond to reality.

Figure 1 shows that for Descartes the world was made up of vortices in contact. The two-dimensional representation of the universe is naturally inadequate, but figures 2, 3, 4 and 5 nevertheless reveal the indisputable existence of huge vortices, the galaxies. So Descartes' concept is founded in realities. Figure 6 shows that the galaxies are not all in the same plane, as Descartes drew them, but that since the universe is three-dimensional, some galaxies are facing us, others are seen in profile and yet others are seen from other angles.

Figures 2, 3, 4 and 5 only represent the luminous centers of the galaxies, and some astronomers have jumped to the conclusion that the galaxies were separated by a vacuum. But invisibility is not the same as vacuum and vacuum is not the same as nothingness. This is a meaningless word, used to hide our ignorance. Since the discovery of radiophony and television, the actual existence of a world capable of generating waves can no longer be disputed. Descartes was right in accepting the continuity of the universe, composed of one and the same substance but varying in aspect, some parts being visible, others invisible, but still capable of moving, therefore dynamic.

Figures 2, 3, 4 and 5 also show that the galaxies are unstable

118

formations whose spiral arms are getting further and further away from the center, to which they will never return. So these spiral nebulae are characterized by their tremendous expulsion of matter and energy, an actual dispersal from the center towards the periphery. The three barred spiral nebulae in figure 6 definitely show the expulsion of matter on the equatorial plane as well as the evolution of galaxies spreading and flattening out through the ages as a result of their rotation.

Since the matter expelled cannot go anywhere, it must be found, as Descartes has drawn it, where the different galaxies meet. Pressed against each other by the movement of galactic expansion, the expelled bodies are compressed so quickly that new galaxies form at the meeting point. The evolution of the universe is thus characterized by the continual dispersal of certain galaxies and the continual formation of new galaxies from the debris of the old ones. The galaxies are thus not all the same age, as the Big Bang theory would have it, for the simple reason that there was no Big Bang consisting of an initial explosion of a compact mass containing all the galaxies. The universe had no beginning but is constantly evolving. The mistake made by the exponents of the Big Bang was to consider as a vacuum, and hence non-existent, the invisible part of the universe which is to be found between the visible center and the periphery where matter is condensed.

Hooke, an eminent successor of Descartes, had the perspicacity to realize that universal substance and movement are two primordial realities which may replace each other. More than two hundred years before Einstein, he postulated the equivalence of mass and energy in a far wider and more accurate form than Einstein, who ignored the mass of interstellar and intergalactic space, whilst for Hooke this space also had mass, since it was capable of generating waves. Hooke thus demonstrated the true motive force of the universe which animates the stars, whose energy derives essentially from the pressure of the surrounding environment.

THE SOLAR SYSTEM

Since the solar system is part of the Milky Way, a spiral nebula in the process of expanding and disintegrating, its evolution is governed by that of the unit to which it belongs. It is therefore wrong to suggest that the solar system was formed through the accretion of a gaseous mass. It is at the level of the galaxy that the compression took place and the solar system has its origin, and is still evolving, in the galactic expansion. It is itself expanding, as shown by Bode's law (figure 7), where the geometric progression in a ratio of 2 means that at the center of the solar vortex (from Mercury to Uranus exactly) there is a centrifugal force proportional to the radius. Over a given period of time, the planet at distance 1 moves 1 to 2, the one at 2 moves on to distance 4, the one at 4 moves on to 8, etc. This means that, as a result of its rotation, the Sun periodically expels a planet.

The same phenomenon of deformation through rotation occurs with the planets, which periodically expel a satellite or quantities of debris in the form of rings when they have become gaseous and scarcely cohere any longer. Figures 8 and 9 clearly show the successive stages of deformation of a rotating sphere up to the expulsion of the part furthest removed from the axis of rotation. The same happens with the stars, which are plastic and cannot escape this phenomenon (see figure 10). The fixed number which has to be added to each element of the geometric progression represents the initial leap made by the ejected mass. Once it has been expelled, the mass is drawn along by the surrounding environment which determines both its travel and its rotation. To understand what takes place at this point it is necessary to study the behavior of vortices. Figure 11 shows that the Earth's orbit is not an ellipse but a spiral, which is demonstrated by the displacement of the aphelion in the direction of travel, producing the so-called anomalistic year, which is 4 minutes and 45 seconds longer than the sidereal year, representing a complete orbit of 360°. Like the photographs of the spiral nebulae, figure 12 shows that a vortex created in the laboratory is denser at the center than at the periphery, whilst figure 13 shows that a vortex engenders in its vicinity another vortex turning the opposite way.

The centrifugal force which prevails in the heart of a vortex means that the external parts (the arms of the spiral nebulae) eventually turn in the opposite direction though constantly tending towards the outside.

In addition, vortices are not just juxtaposed as Descartes drew them but a mass in the heart of a vortex actually begins to spin on its own axis. In fact, the speed of the vortex within which it is spinning decreases from the center towards the periphery (Kepler's law), so that the incorporated mass is drawn along more rapidly on its hemisphere close to the center than on the outside. The diagram in figure 14 shows how a vortex incorporated in another, larger one, evolves.

This theoretical representation has been confirmed by photos showing not only the center of a nebula but also the evolution of the vortex beyond the center (figure 15). Similarly figure 16 shows a nebula which has been more or less crushed by the pressure of the surrounding environment.

When they are expelled from the Sun, the planets, in the course of their life, pass from the center of the solar vortex, which is turning forwards (anti-clockwise), to the periphery, which is turning backwards. Uranus demonstrates how a planet swings from spinning forwards to backwards by turning a somersault, whereby its south pole becomes its north pole (figure 17). Currently Uranus presents one of its poles to the Sun at each solstice. When I was sixteen I spent my holidays in Brugg in the Aargau. The town is on the banks of the Aar, and one of the bridges crosses the river at a point where it is very narrow. When there was a storm, huge whirlpools formed at that point and I loved to watch them. One day, when the river was carrying pieces of wood, I saw one of these pieces spin very rapidly away from the center, submerge suddenly and then reappear further out spinning in the opposite direction. This was such an unexpected sight that a lad standing next to me burst out laughing and looked at me and I did the same. For more than twenty years I did not think about that sight, but when I started thinking about the position of Uranus, I suddenly thought about that piece of wood in Brugg and I understood why Uranus did a somersault. I think I missed the chance of a lifetime by not having a camera at that moment. I still maintain a certain respect for whirlpools (i.e., vortices) because one day when I was swimming in the Aar, quite a way upstream from that tricky point, I was caught up by a whirlpool which left me so exhausted that young Tobler, my hosts' son, helped me back to the bank.

The reversal of the current beyond Uranus explains why the geometric progression is a ratio of 2 stops and gives way to a uniform expansion (1500 million km between Uranus and Neptune and 1500 million km between Neptune and Pluto).

The change of direction of the vortex is also seen in the outermost satellites of the planets of Jupiter, Saturn and Neptune. It is thus a phenomenon common to all vortices which evolve normally. The periodic expulsion of satellites by the planets means that their number increases as the planets get further away from the Sun, but also that the planets lose a considerable part of their mass. After passing through a very condensed environment (the pressure on Venus is 95 atmospheres), they reach an environment which is less and less dense and become gaseous. As their diameter increases, the difference in speed from that of the solar vortex between their faces, diurnal and nocturnal, also increases, which explains why the gaseous planets rotate more quickly. Opponents of the expulsion of the Moon by the Earth put forward the argument of the kinetic momentum, according to which the Earth would in the past have rotated at a speed far greater than there is evidence for. This argument is based on the concept of an Earth which is driven by its own motor, drawing the atmosphere which surrounds it. However, it is obvious that this concept is inadequate and that an acceleration of $g = 9.80$ m/sec of the Earth cannot retain the atmosphere of a planet which is progressing along its orbit at a speed of 30 km/sec. Venus, with its equatorial winds which move above the clouds 60 times faster than the planet is rotating on its own axis, confirms that it is the surrounding environment which determines the movement of the planets, their displacement, their rotation, their state (solid or gaseous) and their deformation, leading to the expulsion of satellites.

The pressure of radiation (figure 19) which drives the comet tails away from the Sun clearly shows that the power of attraction of the stars proportional to their mass is pure fiction. The Sun cannot attract planets and comets by means of its mass and repulse them with its wind. The Bow shock (figure 18) attests that the relation between Sun and Earth is pressure, not attraction.

The displacement of a body implies it's getting closer to some objects and further away from others, but getting closer never implies an essentially attractive force, for which a mechanism has never been discovered. Pressure is the only force governing the universe. The expansion of the solar system and the increasing distance of the planets from the Sun are illustrated by photographs and the measurements of space probes which show that Venus is currently subjected to conditions similar to those experienced by the Earth in the past and that Mars, now capped with ice, once had water, showing that it was at that time closer to the Sun (figure 20).

The photographs taken by NASA of Mercury, the Moon, Phobos and other satellites show that on expulsion innumerable debris accompanied the expelled mass, so that impacts occur far more frequently in the leading hemisphere than in the trailing hemisphere. The photograph of Enceladus taken by NASA (figure 21) shows on the one hand the difference in impacts between the leading and the trailing faces, and on the other hand a tremendous scar, probably due to the concentric pressure of the surrounding environment on the misshapen ejected mass, which then became more or less spherical. As to the close-ups of Phobos taken by NASA, they leave no doubt as to the friction of this satellite against Mars on its expulsion, followed by collisions with the ejected debris (figure 22).

The rings of Saturn are well known (figure 23). Jupiter and Uranus have some too. This cannot be chance and is easily explained by the evolutionary process by which the planets pass through a decreasingly dense environment, in such a way that their cohesion diminishes rapidly. Their satellites also swim in a more and more tenuous environment. Some of them split whilst others, having reached areas where the planetary vortex and the solar vortex differ less and less from each other, are finally drawn in by a movement resulting from the two vortices, and travel through the solar system in the form of comets with very elongated orbits, which bring them into environments of very varying degrees of density, and thus break them up into dust. This is the destiny of planets.

THE PLANET EARTH

Geology as a division of astronomy, offers us close-up images of the stages through which our planet has passed in the course of its evolution in the past.

It is very noticeable that landscapes are characterized by two types of cyclic change in climate. The first type reveals the existence of extremely long periods during which much shorter climatic variations occurred. To understand their significance, one must rid oneself of the preconceived idea that the Earth has always been at the same distance from the Sun and has always completed a full rotation on its axis in 24 hours. All the planets have their own distance and their own period of rotation deriving from their evolution. The Swedish geologist de Geers observed that sediments were characterized by a succession of clear strata, deposited during the period of light and dark strata deposited during the dark period. With his students, he counted the number of "varv," that is, the number of these alternations, in order to determine the length of time which had elapsed since the melting of the previous ice age.

Rotation produces the succession of day and night whilst travel produces the succession of the seasons by virtue of the angle of inclination of the Earth's axis of rotation in relation to its orbital plane. At the poles this angle of inclination produces a succession of six months' day and six months' night. The Grand Canyon in Colorado, well known to all geologists, leaves no doubt as to the superposition of very thick geological strata within which there is a quantity of much finer bands. The photograph in figure 24 is a particularly striking one, providing an irrefutable illustration of the succession of day and night over very long periods, hence the slow rotation of the Earth in ancient times, with some seasonal alternations of light due to the angle of inclination of the axis of rotation. This slow rotation of the Earth with the slow alternation of day and night is also witnessed by the succession of tropical and glacial periods. Figure 25 shows the extent of glaciation in North America in the Quaternary Age, with the exception of the polar region where the ice melted periodically because of the inclination of the axis of rotation. The slow rotation caused the ice to melt over a whole continent when it reached the light. Hence the alternation of emergence and immersion of whole continents observed by geologists. The slow rotation had the effect of determining climates basically by longitude, as shown in figure 26, taken, like most of the others, from my book, *Earth's Flight Beyond*, and as witnessed by the flora and fauna of ancient times. It also meant that different geological periods, held to be successive (which is the case for a given region), were not, for different regions, as illustrated in figure 27, which shows that the Triassic, Jurassic and Cretaceous periods were contemporary in different regions (as today "Noon" only refers to a certain region). Figure 28 explains why a corona of life stretched from one pole to the other across the Equator in primordial times, at dawn and at dusk.

The rotation which was gradually deforming the planet more and more produced the expulsion of the Moon, which was in the area of the Pacific. The waters which flowed into the hollow left by the ejected mass completely altered the dynamic balance of the planet, since their density was three times lower than that of the rock. The continents, which had been stable until then, began to move and drift, as Wegener demonstrated so convincingly (figure 29). The upheavals which occurred at the end of the Secondary Age led some astronomers to wonder whether they were due to a nearby explosion of a supernova. But there is no need to look to the distance, to hypothetical events, to find their cause. The event took place on Earth.

Geologists no longer dispute these continental displacements, as many did at the time of Wegener, and set up a theory of plate tectonics without realizing that the real cause was the expansion and deformation of the terrestrial globe in the context of the expansion of the solar system. The great ridge which divides the Atlantic in two (figure 30) marks the line where Europe and Africa moved away eastwards whilst the American continent drifted west. The volcanoes and the earthquakes around the Pacific are an indication that the movement has not finished (figure 31). The periodic inversions of the magnetism in the zones parallel to the great ridge are due to the slow rotation which meant that this area was sometimes in darkness, sometimes in the light. In *Earth's Flight Beyond* there are further illustrations supplementing this essay on the evolution of our planet, which I have summed up with figure 32.

ATOMS

The structure of atoms and their evolution, characterized by the expansion in the form of radioactivity, are illustrated in *The Universe on the Move*.

LE MONDE DE RENE' DESCARTES,
OU TRAITE' DE LA LUMIERE.

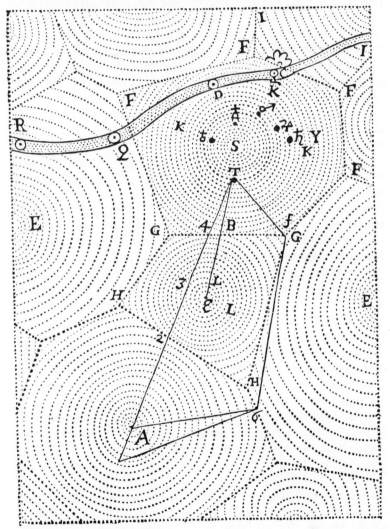

Fig. 1. Descartes' Vortices

Reproduction of a drawing by René Descartes taken from his work Le Monde ou Traité de la Lumière.

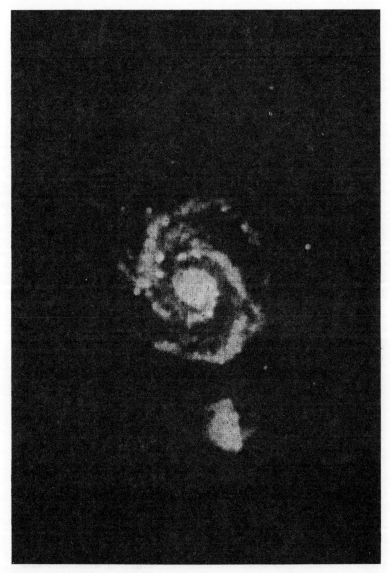

Fig. 2. Descartes Verified

The great stellar systems behave like a vortex. Nebulous spiral M 51 of the Hunting Dogs. Notice the splendid luminous satellites at the extremity of the lower arm. Photo: Mount Wilson.

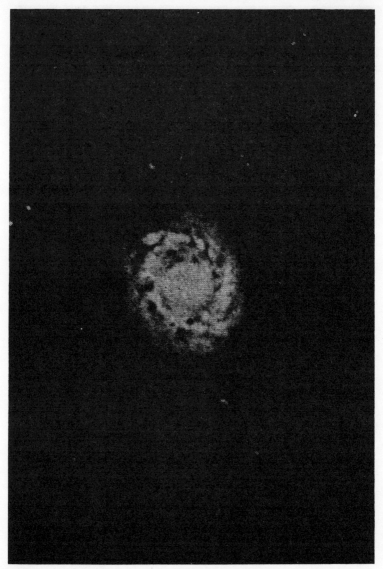

Fig. 3. A Magnificent Vortex

Nebulous spiral NGC 4736 of the Hunting Dogs. One of the best examples of the vortical movement discovered by Descartes. Photo: Mount Wilson.

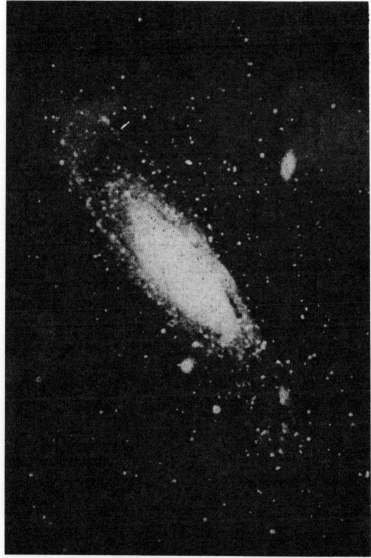

Fig. 4. The Great Nebula of Andromeda

Just as in a whirlwind of dust where the grains which fly about are carried along by the invisible movement of the air, nebulous spirals are a manifestation of the invisible movement of ether. Photo: Yerkes Observatory.

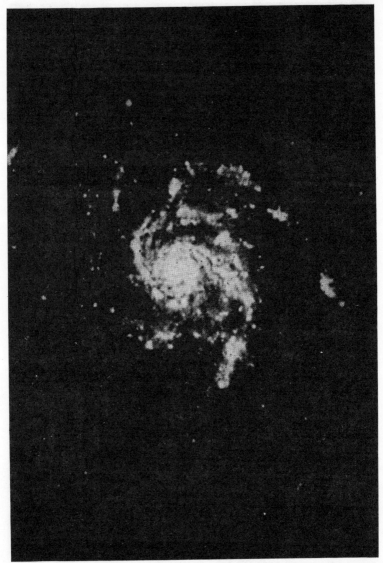

Fig. 5. Nebulous Spiral M 101 of the Big Dipper

The turbulent movement is obvious. More than the wisest of scholarly works such photographs will convince the reader of the superiority of Descartes' vortices over Newton's attraction. Photo: Mount Wilson.

Fig. 6. Barred Nebulous Spirals in Profile

From top to bottom: NGC 4594 in Sextans. NGC 5746 in Virgo, NGC 4565 in Berenice's Hair. The band which crosses the core is an undeniable proof of the existence of interstellar matter. Notice the progressive increase in the development of the arms in relation to the core. These three photographs represent three stages in the progressive flattening of the evolving nebulae. Photo: Mount Wilson.

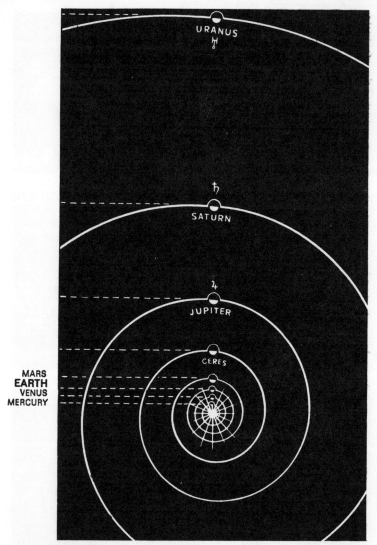

Fig. 7. What Bode's Law Reveals

Bode's law reveals the spiral course of the planets as well as their pro-
gressive movement away from the Sun. Each of the planets successively
occupies the position of Mercury, then of Venus, then of Earth, Mars and
so forth. The relative distances of the planets are given as Mercury = 4,
Venus = 7, Earth = 10, Mars = 16, the Asteroids = 28, Jupiter = 52,
Saturn = 100, Uranus = 196.

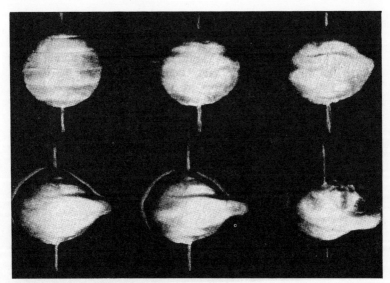

Fig. 8. Deformation of a Rotating Sphere

When a rotating plastic mass swells in the equatorial region, the poles tend to flatten and as the process continues particles come to be thrown off the main mass. Photo: Rochat et Cie, Soleure.

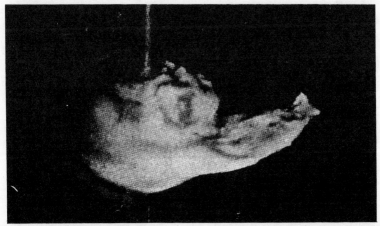

Fig. 9. Imminent Expulsion

Here the action at the Equator is clearly visible. A part of the mass is about to be thrown off. Photo: Rochat et Cie, Soleure.

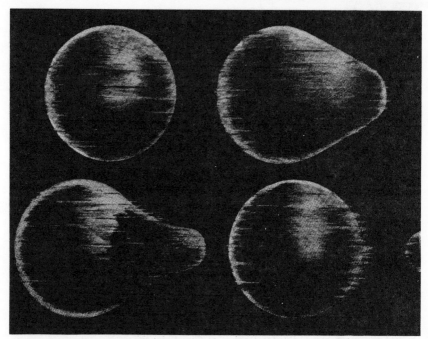

Fig. 10. Birth of a Planet

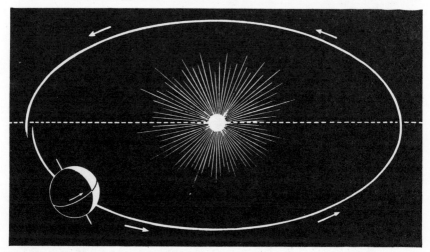

Fig. 11. The Orbit of the Earth

Earth's orbit around the Sun is not a closed ellipse; it is a spiral.

Fig. 12. A Vortex in Laboratory Conditions

The vortices produced in a laboratory are similar to nebulous spirals.
This photograph shows the formation of such a vortex at the tip of an
aircraft wing. Photo: L. Prandtl and O. Tietjens: Hydro-und Aerodynamik,
Springer Verlag, Berlin 1931.

Fig. 13. Two Vortices Turning in Opposite Directions

When the vortices in the preceding picture have been formed, the wing is stopped. The two vortices which then form at the back of the wing turn in opposite directions. Photo: ibid.

Fig. 14. Diagram of a Planetary Vortex and Explanation of Rotation

Less compressed than before, lacking cohesion because of the gaseous layer surrounding it, the planet becomes more and more sensitive to the different pressures acting on its two sides, that which is lit up (closer to the Sun) moving more quickly than the dark side. The heavenly body consequently begins to rotate on its own axis. A planetary vortex is thus formed, made up not only of a solid core but also of a vast circumjacent region. The center (the core and its immediate surroundings) turns in the same direction (positive direction), whereas the two great spiral arms stretching out from it end up by turning back in such a way that the periphery moves in the opposite direction (negative direction).

N.B.—The thin arrows indicate the direction in which the solar vortex is turning. Since the current is slower at the periphery than at the center, this slowing acts like a brake represented by the broad arrows.

Fig. 15. A Complete Vortex

Nebula NGC 7479 in Pegasus. One of the most interesting celestial photographs ever taken. Whereas the majority of the photographs of the nebulae only show the center of the vortex (seen here as the lightest part) with the beginning of the spiral arms, NGC 7479 offers an image of a complete vortex. We can easily discern how the arms stretch outwards in a curve. Photo: Lick Observatory (1908).

Fig. 16. A Crushed Nebula

This photograph of NGC 1300 in Eridanus is also highly instructive because it shows the pressure of the neighboring ether acting on a nebula. It is probable that at the interior of the solar system a number of planetary vortices, particularly in the peripheral zones, undergo similar deformations. Photo: Palomar Observatory.

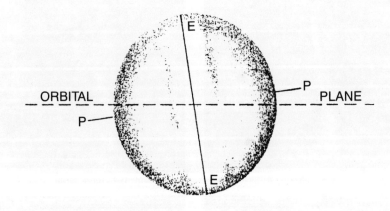

Fig. 17. Uranus' Inclination

Uranus, having arrived in the zone of negative movement, somersaults and advances in its orbit by rolling on its equator which is now almost perpendicular instead of parallel to the orbital plane. The North Pole takes the place of the South Pole by means of a rocking motion. It is in this way that the rotation changes in direction, going from the positive to the negative.

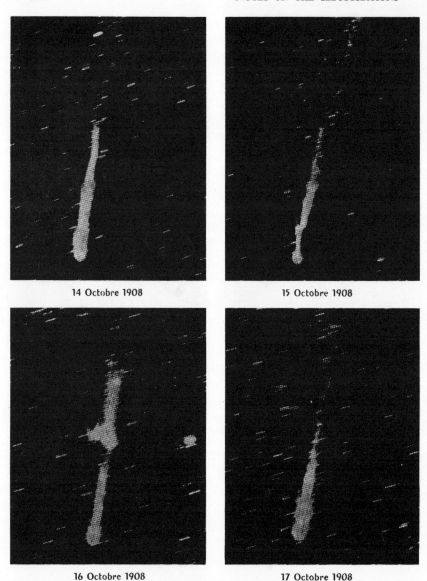

14 Octobre 1908   15 Octobre 1908

16 Octobre 1908   17 Octobre 1908

Fig. 18. The Disintegration of Comets Is Obvious

Photographs of Comet Morehouse taken October 14, 15, 16 and 17, 1908, shows the separation of the tail of the comet. Photo: Yerkes and Juvisy Observatories.

Fig. 19. Day and Night

The succession of light and darkness—irrefutable witness of the slow rotation of the Earth—is clearly visible on this photograph of a region of Greenland. The variations of intensity during the daylight periods are due to the inclination of the Earth in its orbit and allow the determination of the number of revolutions the Earth has made around the Sun in this period. Because such terrains attest to the existence of great millenary cycles as well as to annual variations, they act as excellent calendars. From Riviera polaire, *by E. Hoffer. Kimmerly-Frey Editions géographiques, Berne.*

Fig. 20. Glacial Spread in North America

Note that the extreme north is not ice-covered. This indicates that from time to time the polar region found itself in a light field due to an inclination of the terrestrial axis while the American continent was plunged into darkness. After Daly.

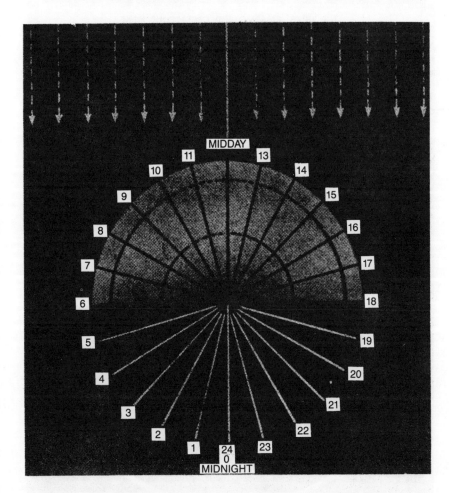

Fig. 21. The Time Zones

The time zones are based on the position of each region in relation to the Sun (longitude). In this way it cannot be noon in Paris and Peking simultaneously.

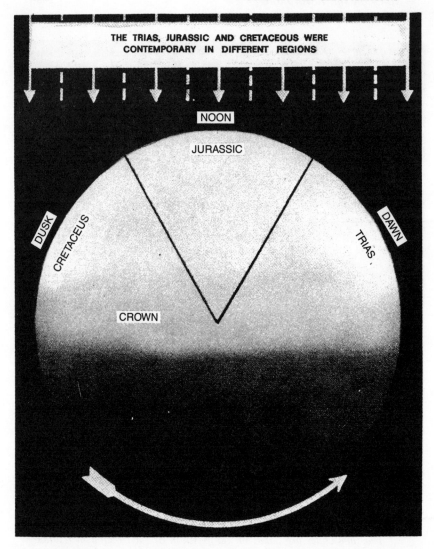

Fig. 22. The Trias, Jurassic and Cretaceous Were Contemporary in Different Zones

When in the Jurassic it was noon in Paris, following the Earth's slow rotation the occidental zone where the Sun rose was still in the Trias, whereas the oriental zone where it set was already in the Cretaceous.

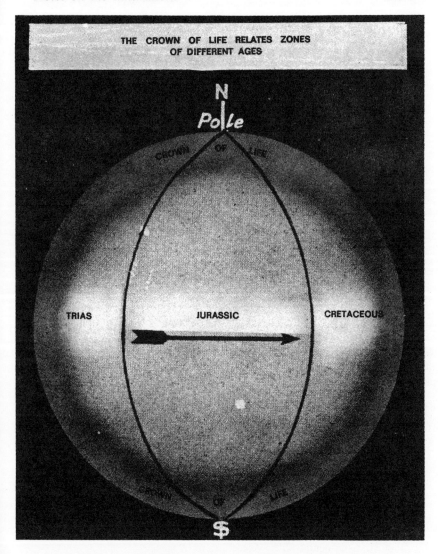

Fig. 23. The Crown of Life Related Zones of Different Ages

The crown of life was a climatic bridge joining the Trias with the Cretaceous, with the Jurassic acting as an intermediary.

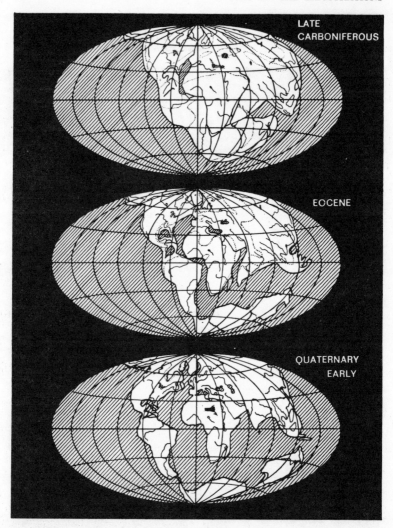

Fig. 24. Continental Shifts According to Wegener

Wegener brilliantly demonstrated that the American continent, earlier attached to Europe and Africa, moved from these in the course of time. The figures represent reconstitutions of the globe at three different geological ages: the late Carboniferous, the Eocene and the late Quaternary. From Wegener: La Genèse des continents et des océans, *Librairie Nizet, Paris.*

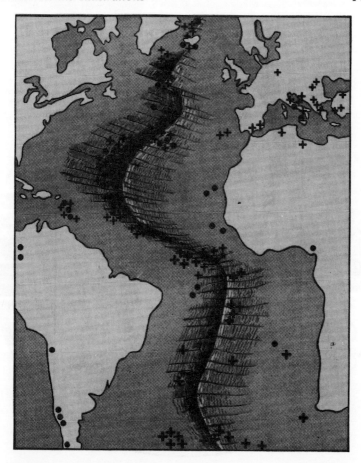

Fig. 25. The Great Ridge, Sites of Earthquakes and Volcanoes in the Atlantic

This map of the bottom of the Atlantic Ocean shows the Great Ridge, the point of departure of the continental shifts. America moved towards the west, Eurasia and Africa towards the east. Unlike the Pacific, where earthquakes and volcanic eruptions mainly take place in its periphery, the Atlantic has a long seismic strip running down its center.

The crosses (+) represent seismic sites, and circles (•) volcanic sites (after Goguel, Traité de tectonique, and recent data about the bottom of the Atlantic).

The faults perpendicular to the Great Ridge are the consequence of the Flatterning of the continual rotating Earth.

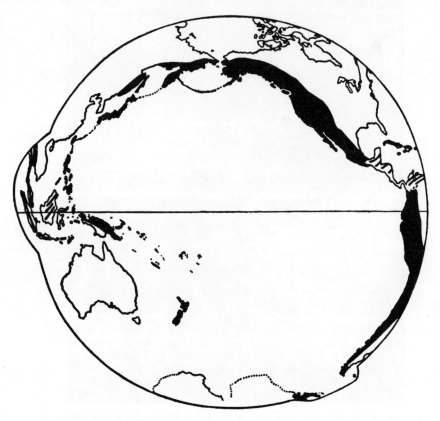

Fig. 26. The Pacific Coast

The recent mountain folds are in black. This fold extending all around the rift is highly suggestive. After Daly: Architecture of the Earth, *D. Appleton Century Co., New York.*

Fig. 27. A Chart of the History of the Earth

In summing up the history of the Earth, we arrive at the following chart, which the reader will kindly read from top to bottom and from left to right, contrary to the geologists' custom of ordering the layers from bottom to top. To our mind the fact that the older layers are usually found under the more recent ones is not a sufficient reason to upset completely the Western way of writing.

Origin: The Earth Is Expelled from the Sun

First Phase: Revolution of the Earth Without Rotation
Archean era: the Earth revolved around the Sun always turning
the same face towards it, as Mercury does now

Second Phase: Very Slow Rotation of the Earth

	Morning	Noon	Evening	Night
	Primary and Algonkian Eras			
1st rotation ·	early Algonkian	middle Algonkian	late Algonkian	late Algonkian early Cambrian
2nd rotation ·	early and middle Cambrian	late Cambrian early Silurien	middle and late Silurian	late Silurian early Devonian
3rd rotation ·	early Devonian	middle Devonian	late Devonian early Carboniferous	middle Carboniferous
4th rotation ·	late Carboniferous early Permian	middle Permian	late Permian	late Permian early Trias
	Secondary Era			
5th rotation ·	middle and late Trias early Jurassic	middle and late Jurassic	late Jurassic early and middle Cretaceous	late Cretaceous Moon's expulsion Dislocation of continents early Eocene
	Tertiary Era			
6th rotation ·	early and middle Eocene	late Eocene Oliogocene	Oligocene Miocene	Miocene-Pliocene (Günz glaciation)

A Chart of the History of the Earth (cont.)

	Quaternary Era: The Appearance of Humanity			
	Interglacial Period:			
	Layer	Man	Industry	Glaciation
7th rotation ·	Sicilian	Pithecanthropus Sinanthropus	Prechellean	Mindel glaciation
8th rotation ·	Tyrrhenian	Heidelberg Man Swanscombe Man	Chellean Acheullian	Riss glaciation
9th rotation ·	Flandrian	Neanderthal Man	Acheuléen Mousterian	Würm glaciation

	Third Phase: Accelerated Rotation of the Earth			
	late Paleolithic	Grimaldi race Cro-Magnon Chancelade *Homo sapiens*	Mousterian Aurignacian Solutrean Magdalenian	
Period of transition	Mesolithic Neolithic		Huts Domestication of animals *Cultivation of cereals Metal Ceramic*	Melting and refreezing of glaciers

| | *Fourth Phase: Daily Rotation of the Earth* | | |
| Historical period | Universal deluge 3500 B.C. Earliest civilizations Historical times | | Abrupt melting of glaciers, then stabilization of climates |

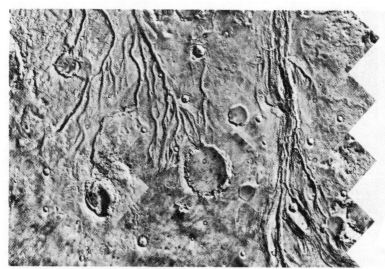

Fig. 28. Martian Rivers

Photo: NASA

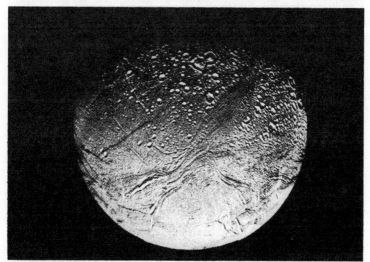

Fig. 29. Enceladus

Satellite of Saturn. Photo: NASA.

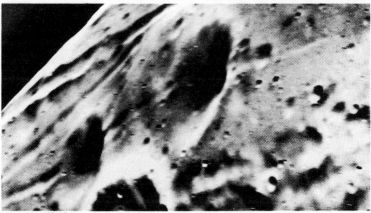

Fig. 30. Phobos

Satellite of Mars. Its surface is covered in grooves attesting to its contact with the planet. Photo: NASA.

Fig. 31. Saturn with Rings and Satellites

This montage combines individual photographs taken by Voyagers One and Two during their Saturn encounters. The satellites are (clockwise from upper right): Titan, Iapetus, Tethys, Mimas, Enceladus, Dione and Rhea. Montage of the Planetary Society. Photo: NASA.